Hormonal and Metabolic Derangements in Renal Failure

Contributions to Nephrology

Vol. 50

Basel · München · Paris · London · New York · New Delhi · Singapore · Tokyo · Sydney

Dialysis Workshop, Grainau, March 21–23, 1985

Hormonal and Metabolic Derangements in Renal Failure

Volume Editors

A. Heidland, Würzburg
E. Quellhorst, Hannoversch-Münden
E. Heidbreder, Würzburg
E. Ritz, Heidelberg
S.G. Massry, Los Angeles, Calif.

64 figures and 23 tables, 1986

 **KARGER**

Basel · München · Paris · London · New York · New Delhi · Singapore · Tokyo · Sydney

Contributions to Nephrology

National Library of Medicine, Cataloging in Publication
    Hormonal and metabolic derangements in renal failure /
    Volume editors, A. Heidland, E. Quellhorst, E. Heidbreder, E. Ritz, S.G. Massry. –
    Basel; New York: Karger, 1986. –
    (Contributions to nephrology; vol. 50) Proceedings of a symposium.
    Includes index.
    1. Endocrine Diseases – complications – congresses 2. Kidney Failure, Acute – complications – congresses
    3. Kidney Failure, Chronic – complications – congresses 4. Metabolic Diseases – complications – congresses
    I. Heidland, August II. Quellhorst, E.A. (Eduard A.) III. Heidbreder, E. IV. Ritz, E. V. Massry, S.G. VI. Series
    W1 C0778UN v. 50
    WJ 342 H812 1986
    ISBN 3-8055-4336-0

Drug Dosage
    The authors and publisher have exerted every effort to ensure that drug selection and dosage set forth in this text are in accord with current recommendations and practice at the time of publication. However, in view of ongoing research, changes in government regulations, and the constant flow of information relating to drug therapy and drug reactions, the reader is urged to check the package insert for each drug for any change in indications and dosage and for added warnings and precautions. This is particularly important when the recommended agent is a new and/or infrequently employed drug.

©   Copyright 1986 by S. Karger AG, P.O. Box, CH-4009 Basel (Switzerland)
    Satz: Satzstudio Frohberg, D-6463 Freigericht
    Printed in Switzerland by Thür AG Offsetdruck, Pratteln
    ISBN 3-8055-4336-0

Contents

Contents

Foreword

During the past decade numerous investigators have elucidated the importance of the kidney within the network of the hormonal regulatory systems. Particularly dialysis treatment in end-stage renal failure gave us the opportunity to investigate its role within the endocrinium and metabolism in humans.

In healthy individuals the kidney is involved in the endocrinium not only by secreting its own hormones such as renin, 1,25-dihydroxycholecalciferol and erythropoietin, but also by the degradation of all polypeptide hormones formed in other glands.

The multiple endocrine dysfunctions in renal failure are only partly caused by the decreased renal and urinary clearances of various hormones. Of further potential importance are: unidentified uremic toxins, deficiency or excess of trace elements, central mechanisms (adrenergic and cholinergic systems, endorphins, adiuretin, prolactin), secondary hyperparathyroidism, intravascular hypervolemia with enhanced secretion of atrial natriuretic peptide and ofther factors. Furthermore various hormone receptors are altered in their function in renal failure. The theme of this symposium will focus to elucidate some of these pathophysiological mechanisms.

In the first presentation Forssmann describes cardiodilatin as a hormone secreted from myoendocrine cells in the auricles of the heart. Further localizations of cardiodilatin are sweat glands, salivary glands and neurons of the central nervous system. According to Heidbreder, the catecholamine concentration in the renal veins is markedly increased as well in acute as in chronic renal failure. Brodde reports about the decreased density and/or responsiveness of α_2- and β_2-adrenoceptors of lymphocytes in uremic patients. During dynamic exercise one can find a considerable rise of β_2-adrenoceptors and a decline of α_2-receptors. Kirschenbaum shows that in progressive renal failure the prostanoid production in the glomerula is markedly enhanced, allowing an increase in SNGFR. Ardaillou calls our attention to the catabolism and renal excretion of antidiuretic hormone in renal insufficiency. Gross reports about the concentration of vasopressin in hyponatremia under these conditions and during hemodialysis. Kaptein adresses herself to thyroid function showing that a low T_4 index in uremia is indicative for hypothyroidism only if the serum TSH level is increased (above 20 μU/ml). Rosenblatt points to the molecular events and regulation of parathyroid hormone secretion with special regard to the role of calcium and other noncalcium modulators.

Massry demonstrates that glucose intolerance in uremic dogs is only observed when parathyroid glands are left intact; parathyroidectomy induces an adequate insulin response. The effects of $1,25(OH)_2$ vitamin D_3 on nonclassical target organs such as endocrine glands (hypophysis, parathyroid, β cells of pancreas, ovary and testis), skeletal muscle, epidermis, vascular smooth muscle and central nervous system are shown by Ritz. Mehls addresses himself to the causes of growth failure in uremia, in spite of generally elevated growth hormone levels. The current knowledge of gonadal function in acute and chronic renal failure and the management of uremic patients with sexual difficulties is reviewed by Kokot and Bommer. Henning reports about neutral steroid metabolites in patients on regular hemodialysis treatment and after renal transplantation. Fyhrquist identifies renin substrate (angiotensinogen) as a possible erythropoietin precursor. Kurtz demonstrates that erythropoietin production in cultures of glomerular mesangial cells can be stimulated by hypoxia or prostaglandins and can be inhibited by indomethacin. The mechanisms of abnormal carbohydrate metabolism in uremia are reviewed by Hörl. Drueke shows that lipid disturbances in experimental renal failure in rats can be corrected by insulin administration.

In order to preserve the authenticity and animation of the discussions of each presentation the editors have tried to reduce their interference to an absolute minimum.

With this symposium a tradition has been abandoned: after five dialysis workshops had been held in Bernried at Lake Starnberg a new residence has been constituted in Grainau/Eibsee without losing the special character of spontaneity and privacy of the meeting. We have to express our gratitude to all participants for their excellent presentations and especially to the Travenol Company for their continuous support of the dialysis workshops.

A. Heidland, E. Quellhorst, E. Heidbreder,
E. Ritz, S.G. Massry

Contr. Nephrol., vol. 50, pp. 1–13 (Karger, Basel 1986)

Cardiac Hormones with Renal Effects

Potential Role in Health and Disease

W.G. Forssmann

Department of Anatomy, University of Heidelberg, Heidelberg, FRG

Recent research has brought a new interesting discovery to light: an endocrine organ has been found in heart tissue and its function is closely related to renal and cardiovascular regulation. The first hint of a specific cell type in the cardiac atria which indicated an endocrine function originated from electron microscopical investigations carried out in the 1950s and 1960s [Kisch, 1956; Bompiani et al., 1959; Jamieson and Palade, 1964; McNutt and Fawcett, 1969; Forssmann and Girardier, 1970]. After Marie et al. [1976] detected a relationship between sodium and water balance and the granule content of the atrial myoendocrine cells in the heart, it became evident for the first time that an endocrine heart function may be implicated in blood volume and electrolyte regulation. DeBold et al. [1981] discovered a diuretic-natriuretic effect of atrial extracts and introduced a diuresis test which was applied as a marker for the isolation of cardiac hormones. Later, a vasorelaxant effect of cardiac atrial extracts was also found [Deth et al., 1982; Currie et al., 1983; Forssmann et al., 1983].

Isolation and Structure of Cardiac Hormones

Various working groups have recently isolated cardiac hormones from porcine, rat, and human atria. The hormones in man are derived from a homologous precursor containing 151 amino acids as determined by cDNA [Yamanaka et al., 1984; Kangawa et al., 1984] and gene studies [Greenberg et al., 1985].

The signal peptide consisting of 25 amino acids is probably processed in the myoendocrine cells forming the 126 amino acid-containing peptide, car-

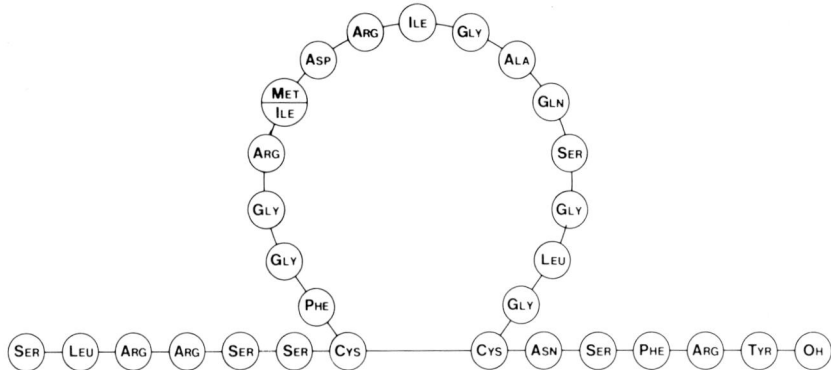

Fig. 1. Biologically active molecule of the C-terminus (CDD 99–126) of porcine cardiodilatin identical to the human α-ANP of Kangawa and Matsuo [1984]. Note the cysteine disulfide bond forming a ring and Met (M) in position 110 which is replaced by Ile in rat and mouse.

diodilatin, which was isolated from porcine atria by our group [Forssmann et al., 1983, 1984]. Smaller molecules which may be posttranslationally processed or artificial fragments of the isolation procedure have been described under various names: cardionatrin [Flynn et al., 1983], atrial natriuretic factor (ANF) [Thibault et al., 1984; Seidah et al., 1984; Misono et al., 1984], atrial natriuretic peptide (ANP) [Kangawa and Matsuo, 1984], atriopeptin [Currie et al., 1984; Geller et al., 1984], and auriculin [Atlas et al., 1984]. Essential for the effectiveness of cardiodilatin is the presence within the molecule of a ring formed by a cysteine bond with methionine in position 110. The latter being replaced by isoleucine in the rat and mouse. The ring formation and the intact methionine are prerequisites for the biological activity (fig. 1). The oxidized methionine-containing peptide, though, exhibits a slight vasorelaxant activity in vitro.

Effects of Cardiodilatin on the Cardiovascular System

The main function of cardiodilatin, which is biologically active in its 126 amino acid form or that processed to smaller C-terminal molecules of cardiodilatin, is evidenced in its role as an antagonist to various vasoconstrictive agents, mainly norepinephrine. This is seen in vitro in various levels of sensitivity on certain vascular smooth muscles. When precontracted with

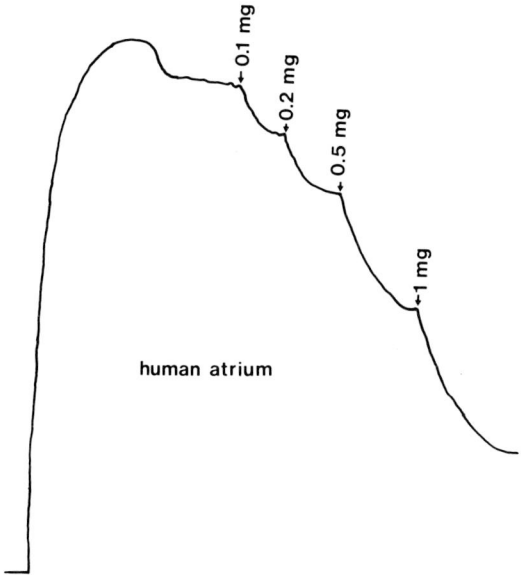

Fig. 2. Dose-dependent relaxation of rabbit aortic smooth muscle precontracted with 10^{-7} M norepinephrine using a crude atrial extract from human right atria.

norepinephrine [see Deth et al., 1982; Currie et al., 1983; Forssmann et al., 1983], cardiodilatin-containing auricular extracts, purified cardiodilatin, or synthetic cardiodilatins produce dose-dependant relaxation (fig. 2). This effect is not observed in mesenterial vessels and is variably sensitive in each segment of the vascular tree. A bolus injection or infusion of cardiodilatin in rats antagonizes the norepinephrine-increased systemic blood pressure (fig. 3). Normal blood pressure is only slightly influenced.

Effects of Cardiodilatin on Renal Function

Kidney function is strongly influenced by cardiodilation. The infusion of cardiodilatin in rats increases the kidney volume by means of an augmented cortical blood flow. The stimulation of renal splanchnic nerves producing a strong reduction of cortical blood flow is antagonized by cardiodilatin. A vasodilation of preglomerular arterial blood vessels and a slight constriction of the vas efferens have been observed by Steinhausen and Forssmann [unpubl.] after injection of natural and synthetic cardiodila-

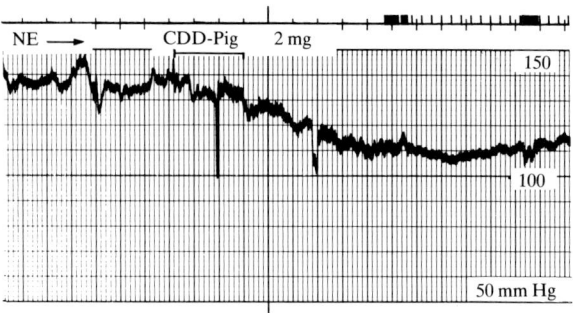

Fig. 3. Influence of prepurified porcine cardiodilatin extract injected into a rat as a bolus. Note the drop of blood pressure, which has previously been raised, by a continuous infusion of norepinephrine.

tin in rats. This is on effective means to produce an increased diuresis-natriuresis because renal tubular ATP is not stimulated by our extracts as already described by Pollock et al. [1983]. Furthermore, no change of tubular transport is observed in isolated perfused nephron segments (experiments carried out with prepurified porcine cardiodilatin by Dr. Greger, Max-Planck-Institut, Frankfurt).

General Role of Cardiac Hormones

Cardiac hormones thus antagonize the effects of norepinephrine released from the hormonal pool such as from the adrenals or from stimulated sympathetic nerves. The secondary effects are a reduction of blood volume by natriuresis-diuresis which in turn results in a decrease in blood pressure. Other long-term influences are observed, such as the reduction of aldosterone synthesis [Atarashi et al., 1984] and a probable inhibition of hormone function of the posterior pituitary gland [Samson, 1985]. This may additionally reduce systemic blood pressure for an extended time.

Release and the Circulating Form of Cardiodilatin

The circulating form of cardiodilatin has been studied. It is greatly increased by volume loading [see Lang et al., 1985]. There is, however, continuous controversy as to whether this is dependent on the cardiac innerva-

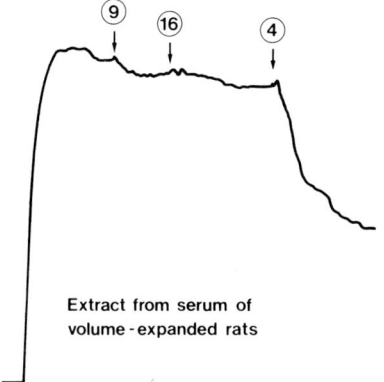

Fig. 4. Biological effect of serum cardiodilatin extracted from a sample of 1 ml rat serum after volume expansion (sample 4). Note the distinct relaxation of the precontracted rabbit aortic smooth muscle. Samples 9 and 16 are from control animals.

tion [Kaczmarczyk et al., 1981]. Its biological activity and chromatographic properties have been determined in our laboratory (fig. 4). According to these findings the molecule is in the range of 3,000–4,000 daltons. Thus, the larger cardiodilatin-126 molecule appears to be processed rapidly from its pools within the myoendocrine cells during hormonal release. The cardiodilatin-126 is therefore considered a prohormone. The hormone found in the circulation is most likely a smaller cardiodilatin peptide which constitutes the C-terminal segment of cardiodilatin-126. Radioimmunoassayable and biologically active cardiodilatin fractions have also been found in human blood.

Morphological Aspects of Cardiodilatin

Cardiodilatin was first detected by immunohistochemistry in porcine, dog, and cat heart by our working group [Metz et al., 1984; Forssmann et al., 1984] and in the rat heart by Cantin et al. [1984] and the human heart by Forssmann and Mutt [1985]. The presence of cardiodilatin-immunoreactivity in human atria is documented in figures 5 and 6 and shown in the area where granules of the myoendocrine cells are seen by electron microscopy (fig. 8).

Further localizations of cardiodilatin have recently been reported: the sweat glands [Reinecke and Forssmann, 1985, unpubl.], the salivary glands

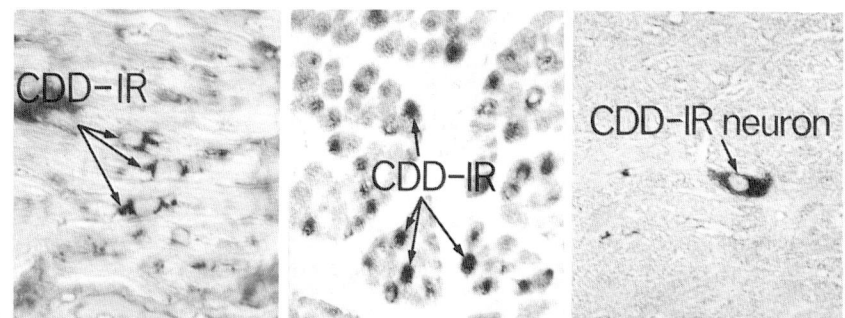

5, 6

CDD-IR

CDD-IR

CDD-IR neuron

7

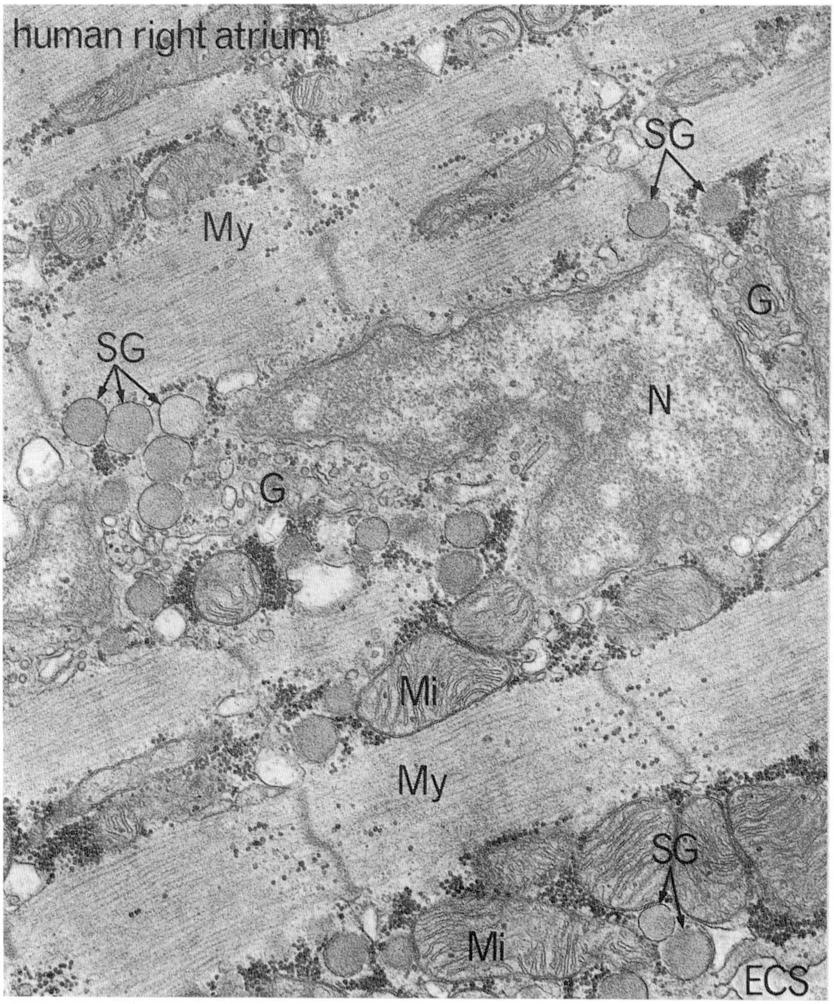

human right atrium

SG

My

G

SG

N

G

Mi

My

Mi

SG

ECS

8

[Cantin et al., 1984] and neurons in the central nervous system [Forssmann and Mutt, 1985]. In the central nervous system, cardiac hormones have also been detected by radioimmunoassay [Tanaka et al., 1984]. The cardiodilatin-containing neurons of the hypothalamus (fig. 7) are located in the region of the cardiovascular modulator centers and other areas of the brain stem. Thus, the brain cardiodilatin-like substance may also be involved in cardiovascular regulation.

Conclusions

The investigations on cardiac hormones carried out so far have partially elucidated: (1) the sites of production of these peptides; (2) the target organs interacting with these substances; (3) the cellular mechanism by which these signal transmitters activate or inactivate the function of the target cells, and (4) the fact that in some genetic disorders these factors are unbalanced.

One of the main target organs for cardiac hormones is the smooth muscle of certain blood vessels, particularly in the kidney within which the blood flow is influenced by preglomerular vasodilation and probably postglomerular vasoconstriction. This results in a tremendous increase of water and sodium excretion. One of the major effects is also to interact (i.e. to inactivate) with the catecholamine-induced changes in renal blood flow. This is demonstrated by the inactivation effects induced by renal nerve stimula-

---

Fig. 5. Cardiodilatin-immunoreactive myoendocrine cells in human right atrium, demonstrated with a C-terminus-specific antibody and the peroxidase-antiperoxidase method. × 300.

Fig. 6. Cardiodilatin-immunoreactive myoendocrine cells in the human right atria seen in a transversely cut plane. The perinuclear zones form a core-like stained area in these sections. × 300.

Fig. 7. Neurons of the hypothalamus (monkey, *Tupaia belangeri*) stained by a C-terminus specific antibody. The neurons are located in the pars magnocellularis of the nucleus periventricularis hypothalami. × 300.

Fig. 8. Myoendocrine cell of human right atrium seen in the electron microscope. The area where cardiodilatin immunoreactivity is seen (see fig. 5) is enlarged and the high amount of specific endocrine granules (SG) is observed. N = Nucleus; My = myofibrils; G = Golgi apparatus; Mi = mitochondria. Some secretory granules are also seen close to the sarcolemma adjacent to the extracellular space (ECS). × 18,000.

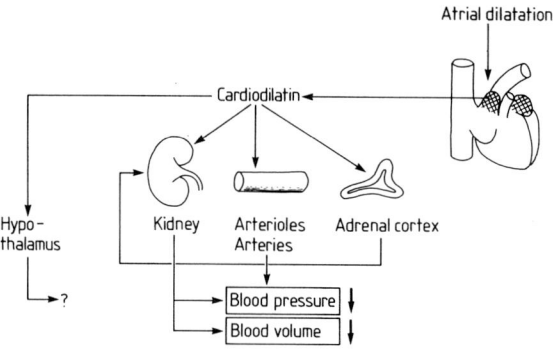

Fig. 9. Schematic drawing of the major functions of cardiodilatin.

tion and norepinephrine infusions. However, we are in the initial phase of comprehending of the clinical importance of cardiac hormones (fig. 9). The implication of cardiac hormones in the diagnosis and therapy of renal and cardiovascular diseases have exponded the field of cardiovascular research.

References

Atarashi, K.; Mulrow, P.J.; Franco-Saenz, R.; Snajdar, R.; Rapp, J.: Inhibition of aldosterone production by an atrial extract. Science *224:* 992–994 (1984).

Atlas, S.A.; Kleinert, H.D.; Camargo, M.J.; Januszewicz, A.; Sealey, J.E.; Laragh, J.H.; Schilling, J.W.; Lewicki, J.A.; Johnson, L.K.; Maack, T.: Purification, sequencing and synthesis of natriuretic and vasoactive rat atrial peptide. Nature, Lond. *309:* 717–719 (1984).

Bompiani, C.D.; Rouiller, C.; Hatt, P.: Le tissue de conduction du cœur chez le rat. Etude au microscope électronique. Archs Mal. Cœur Vaisseaux *52:* 1257–1274 (1959).

Cantin, M.; Gutkowska, J.; Thibault, G.; Milne, R.W.; Ledoux, S.; MinLi, S.; Chapeau, C.: Immunocytochemical localization of atrial natriuretic factor in the heart and salivary glands. Histochemistry *80:* 113–127 (1984).

Currie, M.G.; Geller, D.M.; Cole, B.R.; Boylan, J.G.; YuSheng, W.; Holmberg, S.W.; Neddleman, P.: Bioactive cardiac substances. Potent vasorelaxant activity in mammalian atria. Science *221:* 71–73 (1983).

Currie, M.G.; Geller, D.M.; Cole, B.R.; Siegel, N.R.; Fok, K.F.; Adams, S.P.; Eubanks, S.R.; Gallupi, G.R.; Needleman, P.: Purification and sequence analysis of bioactive atrial peptides (atriopeptins). Science *223:* 67–69 (1984).

DeBold, A.J.; Borenstein, H.B.; Veress, A.T.; Sonnenberg, H.: A rapid and potent natriuretic response to intravenous injection of atrial myocardial extract in rats. Life Sci. *28:* 89–94 (1981).

Deth, R.C.; Wong, K.; Fukozawa, S.; Rocco, R.; Smart, J.L.; Lynch, J.; Awad, R.: Inhibition of rat aorta contractile response by natriuresis-inducing extract of rat atrium. Fed. Proc. *41:* 983 (1982).

Flynn, T.G.; DeBold, M.L.; DeBold, A.J.: The amino acid sequence of an atrial peptide with potent diuretic and natriuretic properties. Biochem. biophys. Res. Commun. *117:* 859–865 (1983).

Forssmann, W.G.; Girardier, L.: A study of the T-system in rat heart. J. Cell Biol. *44:* 1–19 (1970).

Forssmann, W.G.; Hock, D.; Lottspeich, F.; Henschen, A.; Kreye, V.; Christmann, M.; Reinecke, M.; Metz, J.; Mutt, V.: The right auricle of the heart is an endocrine organ. Cardiodilatin as a peptide hormone candidate. Anat. Embryol. *168:* 307–313 (1983).

Forssmann, W.G.; Birr, C.; Carlquist, M.; Christmann, M.; Finke, R.; Henschen, A.; Hock, D.; Kirchheim, H.; Kreye, V.; Lottspeich, F.; Metz, J.; Mutt, V.; Reinecke, M.: The auricular myocardiocytes of the heart constitute an endocrine organ. Characterization of a porcine cardiac peptide hormone, cardiodilatin-126. Cell Tiss. Res. *238:* 425–430 (1984).

Forssmann, W.G.; Hock, D.; Kirchheim, F.; Metz. J.; Mutt, V.; Reinecke, M.: Cardiac hormones: morphological and functional aspects. Clin. exp. Theory Pract. *A6:* 1873–1878 (1984).

Forssmann, W.G.; Mutt, V.: Cardiodilatin-immunoreactive neurons in the hypothalamus of Tupaia. Anat. Embryol. *172:* 1–5 (1985).

Geller, D.M.; Currie, M.G.; Wakitani, K.; Cole, B.R.; Adams, S.P.; Fok, K.F.; Siegel, N.R.; Eubanks, S.R.; Galluppi, G.R.; Neddleman, P.: Atriopeptins. A family of potent biologically active peptides derived from mammalian atria. Biochem. biophys. Res. Commun. *120:* 333–338 (1984).

Greenberg, B.D.; Bencen, G.H.; Seilhamer, J.J.; Lewicki, J.A.; Fiddes, J.C.: Nucleotide sequence of the gene encoding human atrial natriuretic factor precursor. Nature, Lond. *312:* 656–658 (1984).

Jamieson, J.D.; Palade, G.E.: Specific granules in atrial muscle cells. J. Cell Biol. *23:* 151–172 (1964).

Kaczmarczyk, G.; Drake, A.; Eisele, R.; Mohnhaupt, R.; Noble, M.I.M.; Simgen, B.; Stubbs, J.; Reinhardt, H.W.: The role of the cardiac nerves in the regulation of sodium excretion in conscious dogs. Pflügers Arch. *390:* 125–130 (1981).

Kangawa, K.; Matsuo, H.: Purification and complete amino acid sequence of α-human atrial natriuretic polypeptide (α-hANP). Biochem. biophys. Res. Commun. *118:* 131–139 (1984).

Kangawa, K.; Tawaragi, Y.; Oikawa, S.; Mizuno, A.; Sakuragawa, Y.; Nakazato, H.; Fukuda, A.; Minamino, N.; Matsuo, H.: Identification of rat atrial natriuretic polypeptide and characterization of the cDNA encoding its precursor. Nature, Lond. *312:* 152–155 (1984).

Kisch, B.: Electron microscopy of the atrium of the heart. I. Guinea pig. Expl Med. Surg. *14:* 99–112 (1956).

Marie, J.P.; Guillemot, H.; Hatt, P.Y.: Le degré de granulation des cardiocytes auriculaires. Etude planimétrique au cours de différents apports d'eau et de sodium chez le rat. Pathol. Biol. *24:* 549–554 (1976).

McNutt, N.S.; Fawcett, D.W.: The ultrastructure of the cat myocardium. II. Atrial muscle. J. Cell Biol. *42:* 46–67 (1969).

Metz, J.; Mutt, V.; Forssmann, W.G.: Immunohistochemical localization of cardiodilatin in myoendocrine cells of the cardiac atria. Anat. Embryol. *170:* 123−127 (1984).

Misono, K.S.; Grammer, R.T.; Fukumi, H.; Inagami, T.: Rat atrial natriuretic factor. Isolation, structure and biological activities and four major peptides. Biochem. biophys. Res. Commun. *123:* 444−451 (1984).

Pollock, D.M.; Mullins, M.M.; Banks, R.O.: Failure of atrial myocardial extract to inhibit renal Na$^+$, K$^+$-ATPase. Renal Physiol. *6:* 295−299 (1983).

Samson, W.K.: Atrial natriuretic factor inhibits dehydration and hemorrhage-induced vasopressin release. Neuroendocrinology *40:* 277−279 (1985).

Seidah, N.G.; Lazure, C.; Chrétien, M.; Thibault, G.; Garcia, R.; Cantin, M.; Genest, J.; Nutt, R.F.; Brady, S.F.; Lyle, T.A.; Paleveda, W.J.; Colton, C.D.; Ciccarone, T.M.: Amino acid sequence of homologous rat atrial peptides: natriuretic activity of native and synthetic forms. Proc. natn. Acad. Sci. USA *81:* 2640−2644 (1984).

Tanaka, I.; Misono, K.S.; Inagami, T.: Atrial natriuretic factor in rat hypothalamus, atria and plasma. Determination by specific radioimmunoassay. Biochem. biophys. Res. Commun. *124:* 663−668 (1984).

Thibault, G.; Garcia, R.; Cantin, M.; Genest, J.; Lazure, C.; Seidah, N.G.; Chrétien, M.: Primary structure of a high M$_r$ form of rat atrial natriuretic factor. FEBS Lett. *167:* 352−357 (1984).

Yamanaka, M.; Greenberg, B.; Johnson, L.; Seilhamer, J.; Brewer, M.; Friedemann, T.; Miller, J.; Atlas, S.; Laragh, J.; Lewicki, J.; Fiddes, J.: Cloning and sequence analysis of the cDNA for the rat atrial natriuretic factor precursor. Nature, Lond. *309:* 719−722 (1984).

Prof. Dr. W.G. Forssmann, Direktor des Anatomischen Instituts der
Universität Heidelberg, Im Neuenheimer Feld, D−6900 Heidelberg (FRG)

Discussion

Ritz: Thank you Dr. Forssmann, for this brilliant lecture. I heard you 3 weeks ago and it is a testimony to the rapid progression in the field that I just heard some new aspects when listening to you today.

Kokot: You mentioned that you were using immunohistologic investigations? Did you find cardiodilatin also in sweat glands?

Forssmann: Yes, however, the possibility of a crossreacting substance cannot be ruled out.

Kokot: Did you find any correlation between the functional state of the sweat glands and the presence of cardiodilatin?

Forssmann: This has not been investigated so far. About 2 years ago I had postulated that there could be a hormonal substance (cardiosuderin) of the heart which influences sweat secretion. My idea was that if one is sitting in the sauna, cardiac output increases and sweat secretion is enhanced which may be induced by the known mechanism of glandular secretory activity. The hormonal substance to stimulate sweat glands could stem from cardiac myoendocrine cells. We have tried to test sweat gland activity after injecting cardiac extracts in animals. The problem is that our biotest is really not exact enough to talk about results. The biotest we use is to measure sodium excretion by sweat glands and electrical conductance of the skin. Thus, the results we have had with this method are preliminary, but I think there will be demonstrable effects in the future.

Heidland: I have a question concerning the localization of this factor in salivary glands. Is this in the acinar or in the ductal region?

Forssmann: This was described by Cantin et. al. [1984]. We have not described it in the salivary glands and we are not sure whether cardiodilatin is really present there. The pure immunohistochemical finding of Cantin et al. [1984] is that cardiodilatin is located in the acinar cells of the salivary gland. Yet, I think that these immunocytochemical results have to be confirmed by extraction of the peptide in question and its nature has to be established by a biotest and by radioimmunoassay. If the biotest is positive, we may call the substance cardiodilatin-like. It could be that for salivary glands special conditions are required of the extraction procedure. The molecule which is detected by immunocytochemistry may be inactivated by enzymes of the gland during isolation.

Brodde: Is the cyclic GMP increase a trigger mechanism for the release of atrial factor followed by the physiological response? Secondly, are there any hormonal stimuli known which can release cardiodilatin?

Forssmann: At first the effect is very prompt. We have confirmed the results of cyclic GMP increase of other working groups in experiments which were carried out by Dr. Reinecke in our laboratory. The hormonal release mechanism of cardiodilatin is either a neural or a direct stimulus by stretching myoendocrine cells, i.e. a volume load of atria.

Ritz: Just to pursue this cyclic GMP — is this the membrane-bound cyclase?

Forssmann: No, it is a particular cyclase. I believe that cyclic GMP induction triggers the relaxation.

Brodde: This would be the first physiological role of cyclic GMP.

Forssmann: That is very interesting because we have studied rubidium fluxes in vascular smooth muscles treated with various doses of cardiodilatin. Dr. Kreye will present his data in Florence next month. There is apparently no change of rubidium fluxes. Also the membrane potential of smooth muscle cells is not altered when cardiodilatin is added, as seen by intracellular recordings.

Rosenblatt: If I understood you correctly, in most of your studies that you described, you used the entire cardiodilatin molecule of 126 amino acids — the extract of the atria. Is that correct?

Forssmann: Yes, in our experiments we used the natural porcine cardiodilatin-126 and cardiodilatin-88 extracted from right and left atrial appendages.

Rosenblatt: Fine, what I wanted to ask was, is all of the biological activity accountable for, is the 28 amino acid segment that most of the other groups have been using, or have you found any unique biological properties for the amino-terminal portion of the molecule?

Forssmann: I think that the 28 C-terminal segment is sufficient for explaining all effects and that an N-terminal fragment does not qualitatively affect the biological activity.

Rosenblatt: Have you had the opportunity to work with the amino-terminal portion?

Forssmann: Yes, we have — the synthetic N-terminal segments that we have investigated so far do not exhibit the biological activity of cardiodilatin.

Rosenblatt: You described the antagonism of the norepinephrine effect. I was wondering whether you or anyone else in this audience feels that this means that atrial natriuretic factor could eventually be tried in acute renal failure? Up until now, most of the evidence has been that it promotes medullary blood flow.

Forssmann: From the morphological findings it is evident that the corticorenal blood flow is affected by cardiodilatin. You can see it under a microscope in normal rats; an infusion of cardiodilatin increases the cortical blood flow and also the volume of the kidneys. Furthermore, Dr. Steinhausen showed that the vascular tree in the kidney is influenced by cardiodilatin infusions, mainly showing an increase in diameter of the preglomerular arteries. A slight decrease in vas efferens diameter was also observed.

Ritz: It would be unlikely, though, to be helpful in acute renal failure since Reubi showed years ago that a vasodilator does not enhance sodium excretion and improve kidney function.

Koch: How long does it take to exhaust the system if you apply constant stimulus? Or how long does it take to deplete the stores?

Forssmann: Experiments with an accurate volume expansion have been carried out by Lang and co-workers, showing the release of cardiac hormones. From these studies it may be concluded that in acute experiments 10% of the intracardiac cardiodilatin stores are available for release from the atrium. So, there is not a significant loss of the stored pool in an acute volume expansion.

Koch: So it may be a substance involved in chronic volume regulation, not only in acute fluid overload.

Forssmann: In chronic fluid imbalance there is a strong increase or decrease of atrial secretory granules depending upon whether the volume has been expanded or reduced.

Massry: Dr. Forssmann, the way I heard it, one of the major physiological regulator function is apparently the state of fluid and sodium balance. If you expand or deplete it, you have changes in production of cardiodilatin. Has anybody studied the blood levels of this hormone in the state of sodium-retaining conditions, such as congestive heart failure, nephrotic syndrome or cirrhosis of the liver? And if so, is it low or not?

Fyhrquist: In Helsinki we have measured plasma concentrations of atrial natriuretic peptide (ANP) by a specific radioimmunoassay. It appears that in severe congestive heart failure with high atrial pressure we may find very high levels up to 800 pg/ml while we have low plasma ANP levels in essential hypertension and low levels of 5–30 pg/ml in healthy subjects.

Massry: How does that fit with cardiodilatin-regulating sodium balance? I mean, in congestive heart failure, sodium will be retained and we have very high levels of the atrial natriuretic factor. You would expect to loose sodium rather than retain it.

Fyhrquist: We were very surprised that there actually appears to be an escape from the action of the hormone in the state.

Forssmann: I think so, too.

Ritz: There are data in Dr. Lang's laboratory in Heidelberg, acutely you do get an increase in the circulating form. When you chronically change the animal's sodium, it no longer increases.

Forssmann: We have data from different patients which are not yet classified into groups but our values of patients from open heart surgery are between 50 and 300 pg/ml, so your increase to 800 in congestive heart failure is rather tremendous.

Fyhrquist: May I add that we administered ANP by i.v. infusion to ourselves in a dosage of 50 ng/kg/min. We achieved plasma levels very close to those seen in patients with congestive heart failure and we got a natriuresis of about 3-fold of the control levels [Lancet *ii:* 66–69, 1985]. So it looks likely that those would be close to naturally-occurring maximum levels.

Kurtz: The ANF effects on renal function are similar to the blocking of the renin-angiotensin system and there is evidence, at least in rat experiments under in vivo and in vitro conditions, that ANF could inhibit renin release. Have you any evidence from your experiments that cardiodilatin could inhibit renin release of the human kidney?

Forssmann: No, we have not extended our studies in this direction.

Fyhrquist: During ANV infusion in humans we got no effect on the plasma renin activity. We observed lowered aldosterone values as if there would be a block on the aldosterone-stimulating angiotensin II action.

Forssmann: I think in fact that aldosterone production is decreased by cardiac hormones.

Gross: It is intriguing that cardiodilatin seems to be localized primarily in the right atrium because, in states of volume overload, the primary circulatory signals have been shown to stem from the left atrium and from the pulmonary circulation. Would you speculate on this discrepancy?

Forssmann: There is no problem with this because the initial investigations on the morphology of these myoendocrine cells have been carried out with studies on the morphology of the atrial conductive system. That is why people used to investigate primarily the right atrium. The left atrium, however, also contains myoendocrine cells. So the left atrium is very rich in cardiodilatin, too.

Ritz: Thank you, Dr. Forssmann. You have given us a glimpse at an exciting and rapidly expanding area and we are certain to hear about the atrial peptides in the near future.

Contr. Nephrol., vol. 50, pp. 14–27 (Karger, Basel 1986)

Plasma Catecholamines in Renal Failure

E. Heidbreder, K. Schafferhans, R. Götz, K. Bausewein, A. Heidland

Medizinische Universitätsklinik Würzburg, Luitpoldkrankenhaus, Würzburg, BRD

Disorders of the Autonomic Nervous System in Chronic Renal Failure

Methods to directly evaluate the activity of the sympathetic nervous system in human subjects are not available. Only indirect indicators of sympathetic nervous system activity, especially the concentration of plasma norepinephrine, are used in many studies. A reduction of baroreflex sensitivity and pathological Valsalva test results are characteristic symptoms, as well as changes in sweat gland excretion and a decrease in pupillary motor reflex [1].

Restriction of the baroreflex is common in dialysis patients [2–5]. The baroreflex activity was examined in 15 normal patients and 12 hemodialysis patients (8 normotensive, 4 hypertensive) from our dialysis unit with phenylephrine (in the supine position) and nitroglycerine (during orthostasis). In the normal subjects a typical sigmoid curve was clearly discernible whereas in the hemodialysis patients the bunching and distortion of this curve reflected the reduced reflex activity. By calculating the baroreflex sensitivity (heart period/MAP) with phenylephrine the reduction in baroreflex activity in hemodialysis patients compared to normal subjects can clearly be seen (fig. 1).

Renal lesions may alter blood pressure regulation by several mechanisms, e.g. by release of pressor and/or depressor substances. To evaluate the sensitivity of vascular end-organs to exogenous norepinephrine, increasing doses of norepinephrine are infused and the changes in the blood pressure and heart rate registered (fig. 2). The comparison of the heart rate response (in ms) between normal subjects (n = 9) and hemodialysis patients (n = 9), in particular, showed a clearly reduced drop in comparison with the

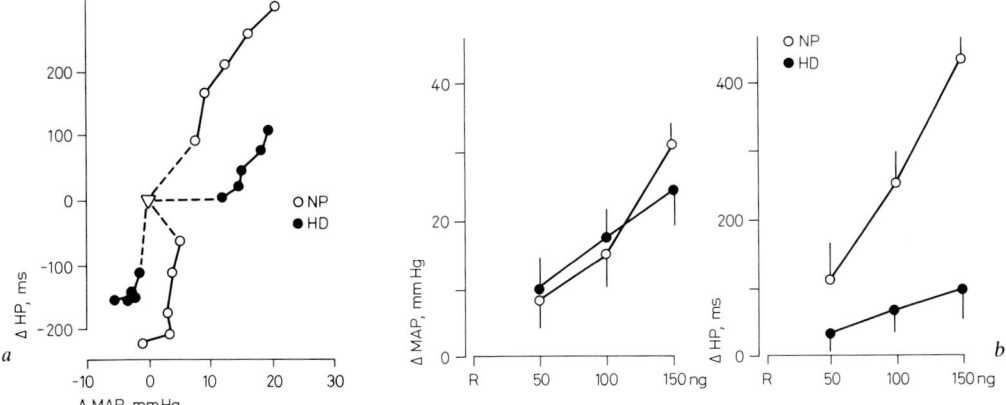

Fig. 1. a Baroreflex activity in 12 patients with end-stage renal failure (solid circles) and 15 normal subjects (open circles). The test results are evaluated by intravenous bolus injection of phenylephrine in resting patients and after sublingual application of nitroglycerine during orthostasis. Triangle = Resting values; NP = normal persons; HD = hemodialysis patients; MAP = mean arterial pressure; HP = heart period. *b* Baroreflex sensitivity in the same persons. The sensitivity is calculated as heart period (ms)-MAP (mm Hg) ratio.

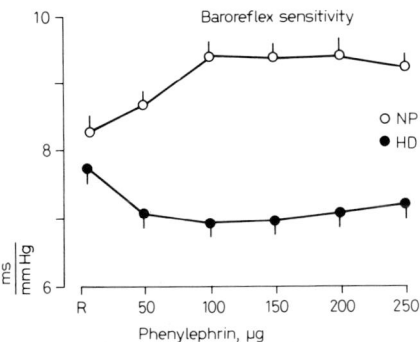

Fig. 2. Cardiovascular pressor Δ-response to norepinephrine. This drug was infused in increasing doses in 9 normal persons (NP) and 9 patients with end-stage renal failure (HD). R = Resting state.

normal subjects. The increase in blood pressure was similar in the two groups. These results confirm tests made by Beretta-Piccoli et al. [6] in mild renal parenchymal disease and by Campese et al. [7] in patients with terminal renal insufficiency. These authors found particularly severe changes in nondialyzed patients and hypothesized that dialyzable material or a down regulation of α-adrenergic receptors of the smooth vascular muscles [8] accompanied by an increase of plasma norepinephrine level are responsible for this phenomenon.

Plasma Catecholamines in Acute and Chronic Renal Failure

In general, the determination of plasma norepinephrine is a parameter, albeit inexact, of the tone of the sympathetic nervous system and as such this method has its limitations [9]. Approximately 80% of the released norepinephrine is inactivated by local re-uptake in the sympathetic varicosities. It is thus not possible to measure the level of norepinephrine released by the sympathetic axon terminals, but rather only that amount which escapes into the circulation.

In the case of chronic renal insufficiency, baseline plasma catecholamines are usually elevated before hemodialysis [7, 10, 11], but in hemodialysis patients they drop or even normalize [12–14]. In a study here, the concentration of catecholamines in the peripheral venous blood in patients suffering from chronic renal insufficiency (n = 21) and acute renal insufficiency (n = 12) was compared (fig. 3). None of the dialysis patients had hypotension. The plasma catecholamines were determined by liquid chromatography (HPLC) using a modification [15] of the Watson [16] method. The norepinephrine values in chronic renal insufficiency patients were generally higher than in the control group and higher in patients not on dialysis than those receiving it. In the case of acute renal failure, the norepinephrine readings are particularly high, almost four times as high as the control group.

The epinephrine readings are less clearly differentiated: in acute renal insufficiency the readings are higher, but not significantly, than the control group. For dopamine no relevant differences were evaluated between the three groups. These results extensively confirm reports by other authors, particularly Campese et al. [17] who hypothesize end-organ resistance to norepinephrine. The mechanisms of these plasma catecholamine changes are discussed in more detail below.

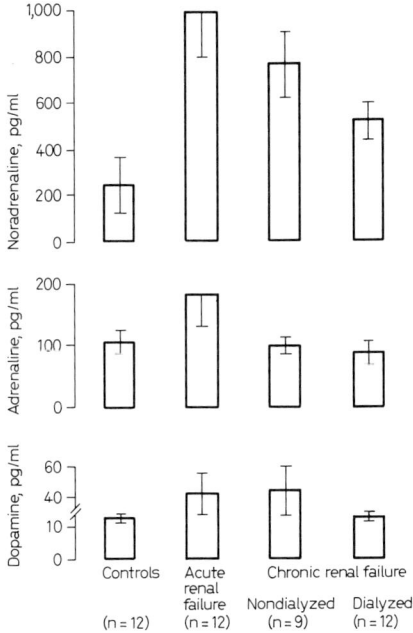

Fig. 3. Plasma catecholamines in normal persons, in patients with acute renal failure and nondialyzed and dialyzed patients with chronic renal failure.

Catecholamine Concentrations at Different Vascular Sites

Several assumptions are involved in the use of plasma norepinephrine as an index of sympathetic neural activity. On the one hand, an increased peripheral vein concentration of norepinephrine reflects a general rise in norepinephrine release, and on the other hand a general increase will necessarily cause an elevation in peripheral venous norepinephrine concentrations. Furthermore, the release of norepinephrine from synaptic clefts into the circulation is uniform in different organs. We therefore tested these assumptions in this study by measuring the arteriovenous difference in norepinephrine across several organs in patients with acute and chronic renal failure and control persons undergoing catheter studies.

Table I. Baseline clinical data in control persons, in patients with acute renal failure (ARF) and in patients with chronic renal failure (CRF)

	Control persons	ARF	CRF
BUN	29 ± 6	95 ± 40	114 ± 41
Creatinine	1.2 ± 0.5	9.3 ± 5	13.7 ± 5
Sodium	140 ± 3	138 ± 6	139 ± 5
pH	7.39 ± 0.04	7.34 ± 0.06	7.27 ± 0.1
Systolic blood pressure	138 ± 8	154 ± 24	156 ± 33
Diastolic blood pressure	85 ± 9	88 ± 10	88.9 ± 9.2
Age	55 ± 8	49 ± 13	51 ± 12

Patients and Methods

A total of 37 persons were examined: 12 control persons, 13 patients with acute renal failure, and 12 patients with chronic renal failure.

The control persons comprised 12 patients with coronary heart disease (8 male, 4 female) who submitted themselves to a heart catheter examination. None of the group had any history, clinical or laboratory indications of kidney or liver malfunction or of the presence of pheochromozytoma. At this juncture we would like to thank Dr. Deeg, Bad Kissingen, for taking the blood samples.

In the patients with acute renal failure all cases, without exception, involved nontraumatically caused conditions: 5 patients had prerenal azotemia, 2 patients postrenal azotemia, 2 renoparenchymal diseases and in the other 4 patients the basic cause of the renal failure was not discernible. Two of the patients died while the kidney condition improved in the other cases.

Twelve patients had terminal chronic renal insufficiency. The examinations were carried out on all patients just before acceptance on the dialysis program. Two of the patients had already been in the chronic hemodialysis program for 3 years.

Table I gives a survey of the most significant clinical data: it shows no relevant difference between patients with acute and chronic renal failure. It should be pointed out that almost all the patients were practically oligo-anuric at the time of the examination.

Most of the patients were catheterized via the femoral vein before hemodialysis with blood being taken (fig. 4) after 30 min rest from the hepatic vein (representative of the mesenteric circulation), the renal and iliac veins (representative of the muscles of the lower extremities) and arterially from the femoral artery. Subsequently, the dialysis catheter was inserted via a wire according to Seldinger's technique.

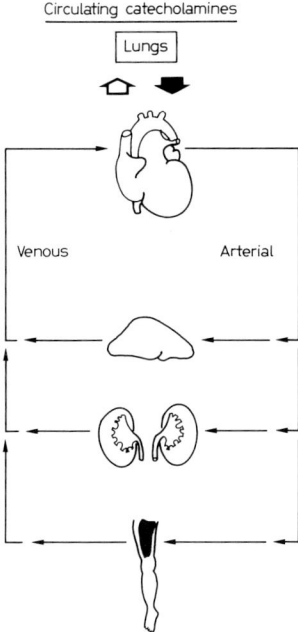

Fig. 4. Catecholamine sampling sites within circulation: hepatic vein, renal vein, iliac vein and femoral artery.

Results

In the acute and chronic renal failure groups the norepinephrine values increased markedly (fig. 5) and these both groups were clearly differentiated from the control group. The highest readings were to be found in the renal venous blood and the lowest relative readings taken from the hepatic vein. Because of the significant individual fluctuations the differences between the arterial and renal venous blood values were only statistically significant in the cases of acute renal failure.

No clear distinction could be made in the comparative epinephrine values (fig. 6): an insignificant increase was discernible only in the renal venous blood of the acute renal failure patients.

The plasma values for dopamine also showed only small deviations (fig. 7) from the control group. The highest values were to be found in the chronic renal failure group and here, too, the highest readings were taken

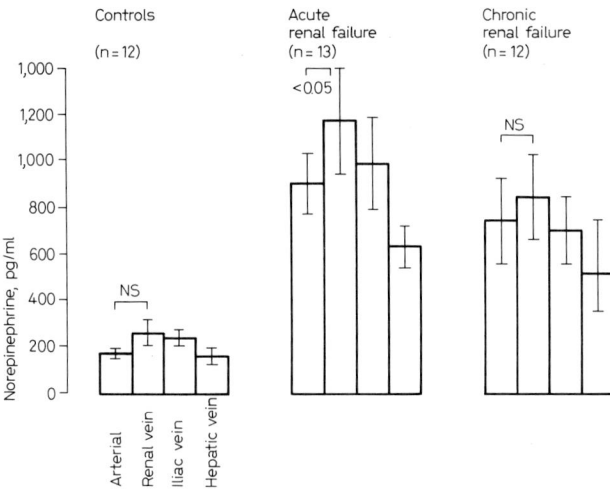

Fig. 5. Plasma norepinephrine concentrations at different sampling sites in control persons, in patients with acute renal failure and in patients with chronic renal failure.

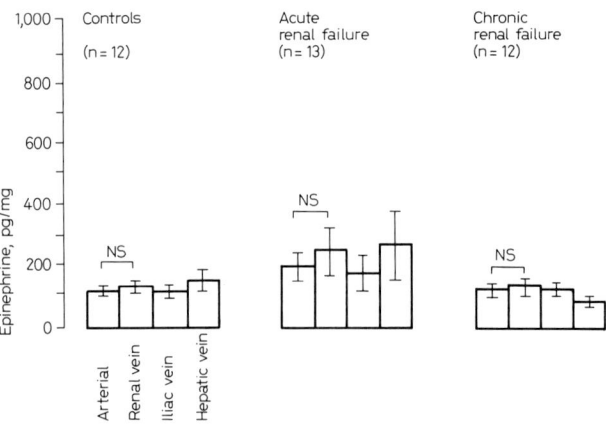

Fig. 6. Plasma epinephrine concentrations at different sampling sites in control persons, in patients with acute renal failure and in patients with chronic renal failure.

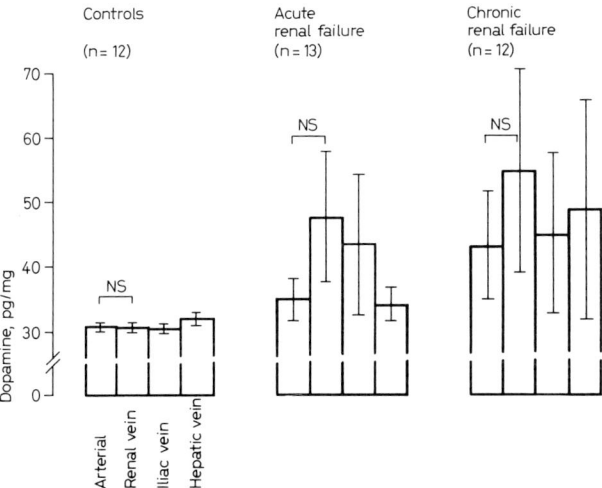

Fig. 7. Plasma dopamine concentrations at different sampling sites in control persons, in patients with acute renal failure and in patients with chronic renal failure.

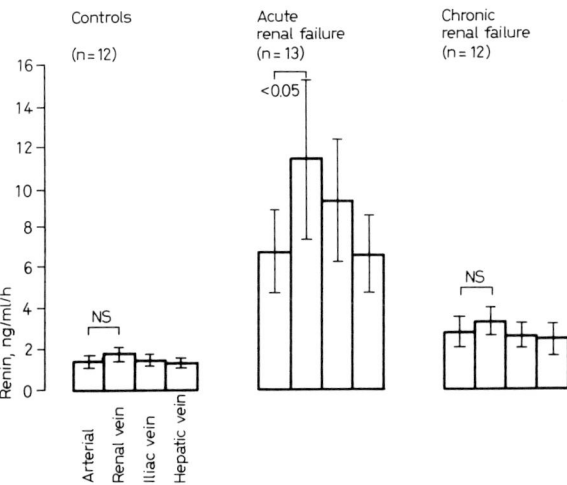

Fig. 8. Plasma renin activity at different sampling sites in control persons, in patients with acute renal failure and in patients with chronic renal failure.

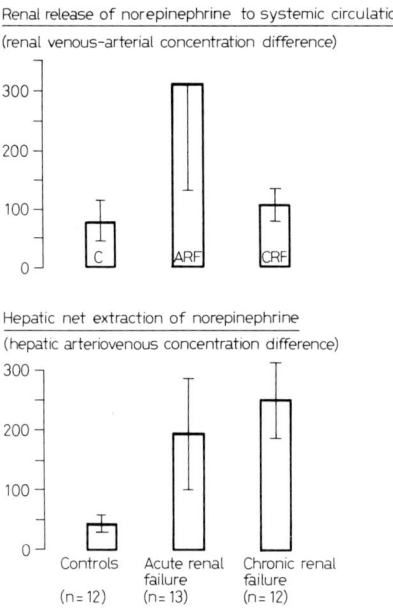

Renal release of norepinephrine to systemic circulation
(renal venous-arterial concentration difference)

Hepatic net extraction of norepinephrine
(hepatic arteriovenous concentration difference)

Fig. 9. Net renal release of norepinephrine into the renal vein and net hepatic extraction of norepinephrine from the circulation in control subjects (C), in patients with acute renal failure (ARF) and in patients with chronic renal failure (CRF).

from the renal venous blood. Statistically significant differences could not be ascertained because of the great interindividual standard variation.

The plasma renin activity (PRA) was also measured (fig. 8) using the radio-enzymatic method described by Sealy and Laragh [18]. The pattern of readings for PRA was very similar to the norepinephrine readings: here, too, the highest readings were found amongst the acute renal failure group and the highest readings within the group taken from the renal venous blood. The norepinephrine and renin values in the renal venous blood of patients suffering from acute renal failure showed a significant correlation (r = 0.80, t < 0.005).

Apart from PRA the only correlations for arterial norepinephrine with other important clinical parameters in the acute renal failure group were the pH and heart rate. All other correlations (blood urea nitrogen, creatinine, systolic blood pressure, diastolic blood pressure, hemoglobin, age) were

Table II. Summary of possible pathogenetic factors in alteration of plasma catecholamines in renal failure

---

Increase in sympathetic nervous system activity
   Reduced baroreceptor sensitivity [19]
   Reduced end-organ response to norepinephrine [7]
   Increased appearance and metabolic clearance norepinephrine [23]
Decrease in renal clearance of catecholamines [21]
Decrease in degradation of catecholamines secondary to reduced COMT activity [21]
Decrease in neuronal uptake of norepinephrine [20]

---

statistically insignificant. The important correlation with urine sodium excretion [14] was impossible to investigate because of the anuria in most patients.

Discussion

The results to this point show that the increased norepinephrine values in chronic renal failure and − even more significantly − in acute renal failure reflect a state of increased sympathetic neural activity. Of particular note (fig. 9) is the increased net renal release of norepinephrine (calculated as net renal venous-arterial concentration difference) into the renal vein, whereas the liver shows an increased net extraction of norepinephrine from the circulation (calculated as net hepatic arteriovenous concentration difference). It is possible that the net release is an indirect indicator of increased renal nervous activity.

From a pathological point of view different mechanisms might come under consideration [19] (table II). Hennemann et al. [20], in their examination of the catabolism of catecholamines in chronic renal failure in rats, found a reduction in neuronal re-uptake and an increase in the catabolism of norepinephrine in the sympathetic nerve endings. Norepinephrine catabolism through catechol-ortho-methyltransferase (COMT) was apparently not altered. Tyrosine-hydroxylase activity was restricted while monoaminoxidase activity (MAO) increased.

A further factor is the reduction in the degradation of catecholamines secondary to the reduced COMT activity in erythrocytes as demonstrated by Atuk et al. [21] in chronic renal failure. These authors hypothesize that reduced enzymatic degradation by this pathway may contribute to the elevated norepinephrine concentration in plasma.

The reduction in renal clearance of norepinephrine in chronic renal failure is a further possible mechanism responsible for the increase in norepinephrine concentrations. The lower values observed by McGrath et al. [12] in 6 anephric patients make it appear unlikely that the reduction or absence of renal clearance plays an important role. The authors hypothesize that the possible effect of angiotensin II might also be a factor in the higher norepinephrine values in non-nephrectomized patients. Hempel et al. [22] showed that O-methylation plays a significant role in the kidneys of dogs.

Izzo et al. [23] observed an increase in the metabolic clearance rate for norepinephrine in chronic renal insufficiency.

Collins et al. [24] showed that the severity of disorders in chronic renal insufficiency is related to secondary hyperparathyroidism in uremia. The change in the MAP in reaction to exogenous norepinephrine is inversely correlated to the blood level of the parathyroid hormone.

How is the net release of norepinephrine from the renal vein to be explained? Is it an index of the overall sympathetic function?

Manhem et al. [25] found no significant differences in normal volunteers between the plasma norepinephrine concentration of the left renal vein, the brachial artery and the peripheral vein.

Oliver et al. [26] noted a baseline overflow of norepinephrine into the renal vein in dogs, this renal overflow showing a significant linear relation to the frequency of the electrical stimulation of the renal nerves. The authors hypothesize that the renal overflow may be used to assess the activity of the renal sympathetic nerves. Kopp et al. [27] also found in dogs that high-level renal nerve stimulation (HLRNS) reduced the renal blood flow by 50% and increased the renal venous outflow of norepinephrine and dopamine, renal uptake of epinephrine being unchanged.

According to this point of view, the present results admit the hypothesis of enhanced renal sympathetic nervous tone in patients with acute and chronic renal failure. More detailed examination of these patients and control groups are necessary, however, to clarify this point.

Enhanced sympathetic neural activity may play an important role in renal hypoperfusion − at least as found in acute renal failure, since in acute renal failure Levitan et al. [28] found reduced plasma norepinephrine levels during stimulation with handgrip and other sympathetic stimuli; changes in plasma norepinephrine concentrations, however, when standing and walking were similar in both hemodialysis and normal subjects [13].

References

1   Röckel, A.; Hennemann, H.; Sternagel-Haase, A.; Heidland, A.: Uremic sympathetic neuropathy after hemodialysis and transplantation. Eur. J. clin. Invest. 9: 23 (1979).
2   Zoccali, C.; Ciccarelli, M.; Maggiore, Q.: Defective reflex control of heart rate in dialysis patients. Evidence for an afferent autonomic lesion. Clin. Sci. 63: 285 (1982).
3   Tomiyama, O.; Shiigai, T.; Ideura, T.; Tomita, K.; Mito, Y.; Shinohara, S.; Tekeuchi, J.: Baroreflex sensitivity in renal failure. Clin. Sci. 58: 21 (1980).
4   Lazarus, J.M.; Hampers, C.L.; Lowrie, E.G.; Merrill, J.P.: Baroreceptor activity in normotensive and hypertensive uremic patients. Circulation 47: 1015 (1973).
5   Wehle, B.; Bevegard, S.; Castenfors, J.; Davidsson, S.; Lindblad, L.E.: Carotid baroreflexes during hemodialysis. Clin. Nephrol. 19: 236 (1983).
6   Beretta-Piccoli, C.; Weidmann, P.; Schiffl, H.; Cottier, C.; Reubi, F.C.: Enhanced cardiovascular pressor reactivity to norepinephrine in mild renal parenchymal disease. Kidney int. 22: 297 (1982).
7   Campese, V.M.; Romoff, M.S.; Levitan, D.; Lane, K.; Massry, S.G.: Mechanisms of autonomic nervous system dysfunction in uremia. Kidney int. 20: 246 (1981).
8   Brodde, O.-E.; Daul, A.: Alpha- and beta-adrenoceptor changes in patients on maintenance hemodialysis. Contr. Nephrol., vol. 41, p. 99 (Karger, Basel 1984).
9   Goldstein, D.S.; McCarty, R.; Polinsky, R.J.; Kopin, I.J.: Relationship between plasma norepinephrine and sympathetic neural activity. Hypertension 5: 552 (1983).
10 Ishii, M.; Ikeda, T.; Takagi, M.; Sugimoto, T.; Atarashi, K.; Igari, T.; Uehara, Y.; Matsuoka, H.; Hirata, Y.; Kimmura, K.; Takeda, T.; Murao, S.: Elevated plasma catecholamines in hypertensives with primary glomerular diseases. Hypertension 5: 545 (1983).
11 Levitan, D.; Massry, S.G.; Romoff, M.; Campese, V.M.: Plasma catecholamines and autonomic nervous system function in patients with early renal insufficiency and hypertension. Effect of clonidine. Nephron 36: 24 (1984).
12 McGrath, B.P.; Ledingham, J.G.G.; Benedict, C.R.: Catecholamines in peripheral venous plasma in patients on chronic haemodialysis. Clin. Sci. mol. Med. 55: 89 (1978).
13 Brecht, H.M.; Ernst, W.; Koch, K.M.: Plasma noradrenaline levels in regular haemodialysis patients. Proc. Eur. Dial. Transplant. Ass. 12: 281 (1975).
14 Lang, R.; Michels, I.; Becker-Berke, R.; Lukowski, K.; Vlaho, V.; Grundmann, R.: Die sympathische Aktivität bei terminaler Niereninsuffizienz und Nierentransplantation. Klin. Wschr. 62: 1025 (1984).
15 Davis, G.C.; Kissinger, P.T.: Strategies for determination of serum or plasma norepinephrine by reverse-phase liquid chromatography. Analyt. Chem. 53: 156 (1981).
16 Watson, E.: Liquid chromatography with electrochemical detection for plasma norepinephrine and epinephrine. Life Sci. 28: 493 (1981).
17 Campese, V.M.; Iseki, K.; Massry, S.G.: Plasma catecholamines and vascular reactivity in uremic and dialysis patients. Contr. Nephrol., vol. 41, p. 90 (Karger, Basel 1984).
18 Sealey, J.E.; Laragh, J.H.: How to do a plasma renin assay. Cardiovasc. Med. 2: 1079 (1977).
19 Lilley, J.L.; Golden, J.; Stone, R.A.: Adrenergic regulation of blood pressure in chronic renal failure. J. clin. Invest. 57: 1190 (1976).
20 Hennemann, H.; Hevendehl, G.; Horler, E.; Heidland, A.: Toxic sympathicopathy in uraemia. Proc. Eur. Dial. Transplant. Ass. 10: 166 (1973).

21 Atuk, N.O.; Bailey, C.J.; Turner, S.; Peach, M.J.; Westervelt, F.B.: Red blood cell cathechol-O-methyl transferase, plasma catecholamines and renal failure. Trans. Am. Soc. artif. internal Organs 22: 195 (1976).
22 Hempel, K.; Lange, H.W.; Kayser, E.F.; Röger, L.; Hennemann, H.; Heidland, A.: Role of O-methylation in the renal excretion of catecholamines in dogs. Arch. Pharmacol. 277: 373 (1973).
23 Izzo, J.L.; Izzo, M.S.; Sterns, R.H.; Freeman, R.B.: Sympathetic nervous system hyperactivity in maintenance hemodialysis patients. Trans. Am. Soc. artif. internal Organs 28: 604 (1982).
24 Collins, J.; Massry, S.G.; Campese, V.M.: Parathyroid hormone and the altered vascular response to norepinephrine in uremia. Am. J. Nephrol. 5: 110 (1985).
25 Manhem, P.; Lecerof, H.; Hökfelt, B.: Plasma catecholamine levels in the coronary sinus, the left renal vein and peripheral vessels in healthy males at rest and during exercise. Acta physiol. scand. 104: 364 (1978).
26 Oliver, J.A.; Pinto, J.; Sciacca, R.R.; Cannon, P.J.: Basal norepinephrine overflow into the renal vein. Effect of renal nerve stimulation. Am. J. Physiol. 239: f371 (1980).
27 Kopp, K.; Bradley, T.; Hjemdahl, P.: Renal venous outflow and urinary excretion of norepinephrine, epinephrine, and dopamine during graded renal nerve stimulation. Am. J. Physiol. 244: E52 (1983).
28 Levitan, D.; Massry, S.G.; Romoff, M.S.; Campese, V.M.: Autonomic nervous system dysfunction in patients with acute renal failure. Am. J. Nephrol. 2: 213 (1982).

Prof. Dr. E. Heidbreder, Medizinische Universitätsklinik, Abteilung Nephrologie, Josef-Schneider-Strasse 2, D−8700 Würzburg (FRG)

Discussion

Ritz: Thank, you, Dr. Heidbreder, for this overview and these beautiful findings on intrarenal sympathetic activity.

Brodde: I have one question regading your norepinephrine experiments on blood pressure in healthy volunteers and in dialysis patients. You found no difference in the blood pressure response. I would like to ask you, what was the blood pressure of your dialysis patients? Were they normotensive, hypertensive or were they hypotensive?

Heidbreder: All the patients were normotensive.

Brodde: Okay, if they were normotensive, there should be no difference.

Ritz: There has been a report by McGrath that he found a correlation between circulating epinephrine levels and blood pressure in dialysis patients, while he failed to find a correlation between norepinephrine levels and blood pressure. This, of course, would go along nicely with the Brown theory of presynaptic amplification. Can you explain the difference with your results? Probably you failed to find a correlation between epinephrine levels and blood pressure in your chronic renal failure patients.

Heidbreder: We couldn't find any correlation with epinephrine but we found a correlation with norepinephrine similar to Massry's group.

Massry: These are really beautiful data and I enjoyed listening to them. I wonder why you dismiss the role of excess PTH in acute renal failure. You know, it's a state of secondary hyperparathyroidism, and I wonder whether you measured the blood level of PTH and whether there is a correlation between the blood levels of norepinephrine and PTH in acute renal failure. In parathyroidectomized rats and normocalcemia, for example with a state of chronic renal failure of 3 weeks, the levels of norepinephrine are normal, while in animals with the same severity of chronic renal failure with intact parathyroid glands, the blood levels of norepinephrine are high. So I propose to go back and measure the PTH in the blood of the acute renal failure patients. Probably, you will find a relationship. I say this because Dr. Weidmann and others have shown that the blood levels of the catecholamines are calcium-dependent. As you know, parathyroid hormone may act as an ionophore in many systems. It is possible that through the high levels of PTH you can stimulate the production of norepinephrine. This has been shown to be true with prostaglandins.

Ritz: Apparently, Dr. Heidbreder, you speculate in document with your data that there is a state of increased sympathetic tone within the kidney in acute renal failure. Of course, the obvious question is, does this contribute to the maintenance phase of acute renal failure and did you try pharmacological interventions? Does this improve renal blood flow if you do alpha-blockade, for instance? Is there any information on this? If I am not mistaken, the Bern group (Dr. Weidmann) did it and did not improve renal blood flow with alpha-blockade, at least.

Heidbreder: No, we didn't perform such investigations as yet.

Contr. Nephrol., vol. 50, pp. 28–35 (Karger, Basel 1986)

Impaired Regulation of α- and β-Adrenoceptor Function in Chronic Renal Insufficiency[1]

Otto-Erich Brodde, Anton Daul

Biochemical Research Laboratory, Medical Clinic and Policlinic,
Division of Renal and Hypertensive Diseases, University of Essen, Essen, FRG

During haemodialysis treatment of chronic renal failure, disturbances of the sympathetic nervous system activity have often been observed. A defective function of sweat glands, reduced elevation of blood pressure in response to sustained hand grip exercise, non-volume-responsive chronic hypotension [9], elevated plasma noradrenaline levels [2], and reduced responsiveness to noradrenaline infusion [1] have been reported. We have recently shown that such end-organ resistance to adrenergic stimulation in haemodialysis patients might be due to a reduced responsiveness of α- and β-adrenoceptors [5] since in platelets of haemodialysis patients adrenaline-induced inhibition of adenylate cyclase activity (via α_2-adrenoceptor stimulation) was significantly reduced and in lymphocytes cyclic AMP responses to isoprenaline (via β_2-adrenoceptor stimulation) were significantly diminished. The aim of the present study was to further evaluate changes of the sympathetic activity in patients on chronic haemodialysis treatment. For this purpose, we have determined the effects of acute stimulation of the sympathetic activity by dynamic exercise on a bicycle ergometer on the densities of platelet α_2-adrenoceptors (assessed by ^{3}H-yohimbine binding) and of lymphocyte β_2-adrenoceptors (assessed by $(-)-^{125}$iodocyanopindolol [ICYP] binding) in normotensive dialysis patients and in dialysis patients with interdialytic hypotension and compared with those in healthy controls.

[1] This work was supported by the Landesamt für Forschung Nordrhein-Westfalen and the Sandoz-Stiftung für therapeutische Forschung.

Subjects and Methods

Seven male healthy volunteers (mean age: 25.6 ± 2.2 [19−27] years; mean blood pressure: $120 \pm 3/75.5 \pm 2.4$ mm Hg), 4 normotensive dialysis patients (3 males, 1 female; mean age: 34 ± 3.5 [25−42] years; mean blood pressure: $124 \pm 4/79 \pm 3.3$ mm Hg) and 4 male dialysis patients with interdialytic hypotension (mean age: 32 ± 2.2 [27−37] years; mean blood pressure: $97 \pm 10/61 \pm 5$ mm Hg) participated in the study after having given informed written consent. Exercise was carried out in a quiet air-conditioned room always between 10 and 12 a.m., i.e. 10−20 h after the last dialysis. The subjects assumed the supine position and a cannula was inserted into an antecubital vein. After 1 h of rest exercise was performed on a bicycle ergometer (Bosch, Berlin) in a supine position starting with an initial work load of 25 W in haemodialysis patients or 50 W in controls. Work load was increased by 25 W every 2 min until 80% of the maximal heart rate (200−age) was reached. The final work load (75−125 W in the haemodialysis patients, 100−150 W in the controls, respectively) was kept constant until a total exercising time of 15 min was reached. Blood samples were obtained immediately prior to exercise, at the end of exercise and 1 h after exercise. Blood pressure and heart rate were recorded automatically by a Tonomed (Speidel und Keller, Jungingen, FRG) and by an electrocardiogram.

ICYP binding to lymphocyte β_2-adrenoceptors and ^{3}H-yohimbine binding to platelet α_2-adrenoceptors was performed as recently decribed [6, 7]. Plasma catecholamines were assessed by an HPLC method with electrochemical detection.

The experimental data given in the text and figures are means $\pm$ SEM of n experiments.

Results

In healthy volunteers dynamic exercise for 15 min on a bicycle in supine position led to an increase of systolic blood pressure from 120 to 185 mm Hg; concomitantly, plasma catecholamines rose from 0.35 to 0.73 ng/ml (noradrenaline) and 0.05 to 0.11 ng/ml (adrenaline), respectively (fig. 1). β_2-Adrenoceptor density in lymphocytes increased from $1,066 \pm 67$ up to $2,011 \pm 144$ ICYP binding sites/cell (n = 7), while platelet α_2-adrenoceptors decreased from 300 ± 39 to 264 ± 20 fmol ^{3}H-yohimbine bound/mg protein (n = 7, fig. 1). One hour after exercise β_2-adrenoceptor density, systolic blood pressure and plasma catecholamines had reached pre-exercise values, while α_2-adrenoceptor density in platelets further decreased by about 10% (fig. 1).

The increase in lymphocyte β_2-adrenoceptor density seems to be mediated by β-adrenoceptor stimulation, since pre-treatment of the volunteers with propranolol (5 mg i.v. 45 min prior to exercise) markedly attenuated exercise-induced increase in β_2-adrenoceptor density (fig. 2).

In normotensive patients on chronic haemodialysis treatment exercise-induced α_2- and β_2-adrenoceptor changes were very similar to those in

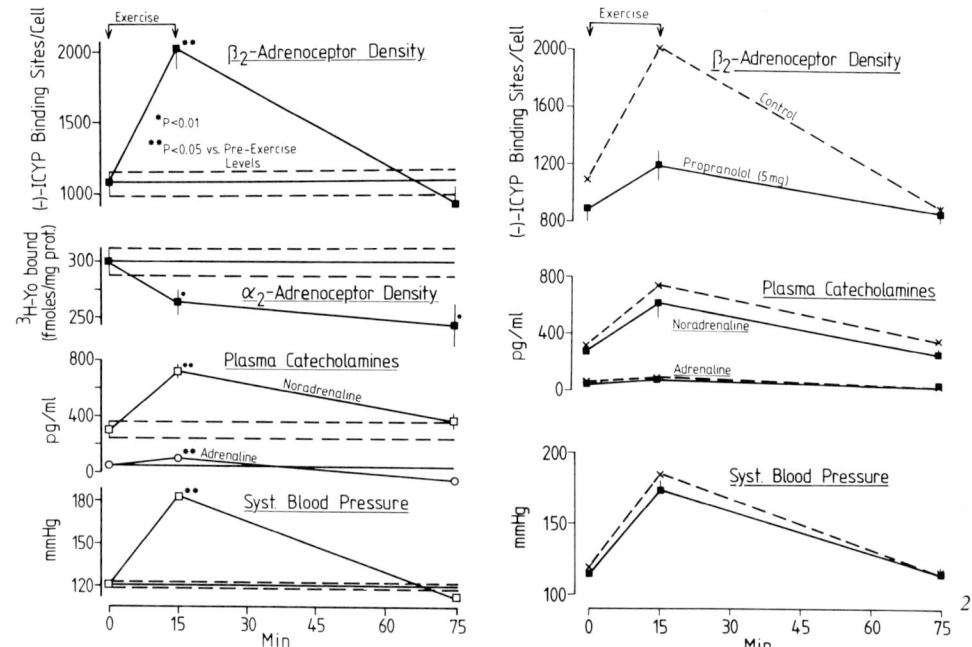

Fig. 1. Dynamic exercise-induced changes (15 min on a bicycle at 80% of maximum heart rate) in lymphocyte β₂-adrenoceptor density, platelet α₂-adrenoceptor density, plasma catecholamine concentrations and systolic blood pressure in 7 healthy volunteers.

Ordinates (from top to bottom): β₂-adrenoceptor density in specific ICYP binding sites/cell; α₂-adrenoceptor density in fmol ³H-yohimbine specifically bound/mg protein; plasma catecholamine levels in pg/ml and systolic blood pressure in mm Hg. Abscissa: time in minutes. Given are means ± SEM. Solid horizontal lines and broken lines = Pre-exercise values ± SEM, determined after 1 h of rest in the supine position.

Fig. 2. Effects of propranolol-induced changes (5 mg i.v. 45 min before exercise) on dynamic exercise (15 min on a bicycle at 80% of maximum heart rate) in lymphocyte β₂-adrenoceptor density, plasma catecholamine concentrations and systolic blood pressure in 7 healthy volunteers. For details see legend to figure 1. For comparison, control responses in the absence of propranolol from figure 1 are given in broken lines. Given are means ± SEM.

healthy controls. In addition, α₂- and β₂-adrenoceptor densities in these patients were not significantly different from control (data not shown). On the contrary, in the 4 patients with interdialytic hypotension platelet α₂-adrenoceptor and lymphocyte β₂-adrenoceptor densities were significantly lower than in control (fig. 3). In these patients, dynamic exercise induced only a

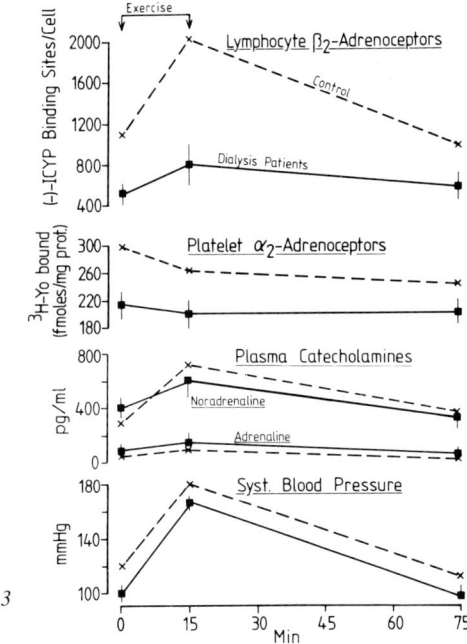

Fig. 3. Dynamic exercise-induced changes (15 min on a bicycle at 80% of maximum heart rate) in lymphocyte β_2-adrenoceptor density, platelet α_2-adrenoceptor density, plasma catecholamine concentrations and systolic blood pressure in 4 hypotensive dialysis patients. For details see legend to figure 1. The control responses in healthy volunteers from figure 1 are given in broken lines. Given are means ± SEM.

small increase in β_2-adrenoceptor density by approximately 25%, while platelet α_2-adrenoceptor density was not affected by exercise.

Discussion

During chronic haemodialysis treatment patients often become hypotensive and insensitive to adrenergic stimulation [1, 9]. In these patients the density and/or responsiveness of α_2- and β_2-adrenoceptors in circulating blood cells is decreased; in addition, peripheral α_1-adrenoceptor responsiveness (as measured by blood pressure responses to phenylephrine) and β_1-adrenoceptor responsiveness (as measured by the effects of isoprenaline on plasma renin activity) is reduced [5]. While the diminished β-adrenoceptor responsiveness appears to be due to a postreceptor defect (impairment of adenylate cyclase activity) [5], the decreased α-adrenoceptor responsiveness seems to be caused by down-regulation through endogenous catecholamines, which are elevated in haemodialysis patients [2]. Thus, changes in catecholamines and adrenergic receptors in hypotensive dialysis patients resemble those in patients with phaeochromocytoma (with high plasma catecholamines and low platelet α_2-adrenoceptor density) [4], but are in sharp contrast to those in patients with asympathicotonic orthostatic hypotension, where recently low catecholamines and a high α_2-adrenoceptor density and responsiveness were shown [3]. Accordingly, yohimbine, which has been found to be of beneficial effect in asympathicotonic orthostatic hypotension [3], cannot be used to improve arterial hypotension in chronic dialysis patients.

The results of the present study demonstrate that not only long-term, but also acute regulation of adrenergic responsiveness is impaired in hypotensive dialysis patients. As shown in figure 1, acute stimulation of the sympathetic activity by dynamic exercise on a bicycle leads in healthy volunteers to an about 100% increase in β_2-adrenoceptor density in lymphocytes, in accordance with recently reported data from several groups [6, 8, 10]. The α_2-adrenoceptor density, on the contrary, was reduced by about 15–20% presumably due to endogenous down-regulation by increased plasma noradrenaline levels. The mechanism of the acute agonist-induced 'up-regulation' of lymphocyte β_2-adrenoceptors is not known. However, one possible explanation may be taken from the recent observation that in rat reticulocyte ghosts stimulation of β-adrenoceptors increases enzymatic methylation of phospholipids, which may unmask 'cryptic' β-adrenoceptors resulting in an increased receptor number [11]. The fact that the exercise-induced increase in lymphocyte β_2-adrenoceptor density can be prevented by pre-treatment with propranolol, supports the view that indeed β-adrenoceptors are involved in this process.

In hypotensive dialysis patients, on the contrary, the exercise-induced increase in lymphocyte β_2-adrenoceptor density was significantly lower than in controls. In addition, platelet α_2-adrenoceptor density was not changed during or after exercise. These results, therefore, favour the idea that in hypotensive dialysis patients not only long-term but also acute regulation of α- and β-adrenoceptors is impaired. Such an impaired regulation of adrenergic receptors – the targets of the catecholamines – may contribute to the disturbances of the sympathetic nervous activity in patients on chronic haemodialysis leading to arterial hypotension.

References

1   Botey, A.; Gaya, J.; Montoliu, J.; Torras, A.; Rivera, F.; Lopez-Pedret, J.; Revert, L.:
    Postsynaptic adrenergic unresponsiveness in hypotensive haemodialysis patients. Proc.
    Eur. Dial. Transplant Ass. *18:* 586–591 (1981).

2   Brecht, H.M.; Ernst, W.; Koch, K.M.: Plasma noradrenaline levels in regular
    hemodialysis patients. Proc. Eur. Dial. Transplant Ass. *12:* 281–289 (1975).

3   Brodde, O.-E.; Anlauf, M.; Arroyo, J.; Wagner, R.; Weber, F.; Bock, K.D.: Hypersen-
    sitivity of adrenergic receptors and blood-pressure response to oral yohimbine in ortho-
    static hypotension. New Engl. J. Med. *308:* 1033–1034 (1983).

4   Brodde, O.-E.; Bock, K.D.: Changes in platelet alpha$_2$-adrenoceptors in human
    phaeochromocytoma. Eur. J. clin. Pharmacol. *26:* 265–267 (1984).

5   Brodde, O.-E.; Daul, A.: Alpha- and beta-adrenoceptor changes in patients on mainte-
    nance hemodialysis. Contr. Nephrol., vol. 41, pp. 99–107 (Karger, Basel 1984).

6   Brodde, O.-E.; Daul, A.; O'Hara, N.: β-Adrenoceptor changes in human lymphocytes,
    induced by dynamic exercise. Arch. Pharmacol. *328:* 362 (1985).

7   Brodde, O.-E.; Hardung, A.; Ebel, H.; Bock, K.D.: GTP regulates binding of agonists
    to α_2-adrenergic receptors in human platelets. Archs int. Pharmacodyn. Thér. *258:*
    193–207 (1982).

8   Butler, J.; Kelly, J.G.; O'Malley, K., Pidgeon, F.: β-Adrenoceptor adaption to acute
    exercise. J. Physiol., Lond. *344:* 113–118 (1983).

9   Campese, V.M.; Romoff, M.S.; Levitan, D.; Lane, K.; Massry, S.G.: Mechanisms of
    autonomic nervous system dysfunction in uremia. Kidney int. *20:* 246–253 (1981).

10 Middeke, M.; Remien, J.; Holzgreve, H.: The influence of sex, age, blood pressure and
    physical stress on β$_2$-adrenoceptor density in mononuclear cells. J. Hypertens. *2:*
    261–264 (1984).

11 Strittmatter, W.J.; Hirata, F.; Axelrod, J.: Regulation of the β-adrenergic receptor by
    methylation of membrane phospholipids; in Dumont, Greengard, Robison, Adv. Cyclic
    Nucleotide Res., vol. 14, pp. 83–91 (Raven Press, New York 1981).

Prof. Dr. O.-E. Brodde, Nephrologisches Labor der Medizinischen Klinik,
Abteilung für Nieren- und Hochdruckkranke, Universitätsklinikum,
Hufelandstrasse 55, D–4300 Essen 1 (FRG)

Discussion

Ritz: Thank you, Dr. Brodde, for this beautiful overview of the issue of catecholaminergic receptors which has given us a new handle on understanding the sympathetic abnormalities
in uremic patients.

Let me start by asking you one question. You showed data on α_1-adrenergic responsiveness using phenylephrine. You plotted on the ordinate the absolute increment of blood pressure and pointed out that such an increment was less in patients who were hypotensive. Does
this still hold true when you give your data as percent blood pressure increase? Of course, an
absolute increase of 10 mm Hg is proportionately less for a patient who has a basal pressure of
100 mm Hg than for one who has 80 mm Hg.

Brodde/Daul

34

Brodde: If I calculate percent blood pressure increase, there is still a decrease in responsiveness in hypotensive patients. In addition, not only is the increase of blood pressure decreased but also the dose of phenylephrine necessary to increase blood pressure is much higher in hypotensive than in normotensive patients.

Ritz: So, if I understand you correctly, both the absolute and the fractional increase is diminished. This would eliminate my objection.

Massry: I may not have understood you well but in the experiments with the exercise when you blocked the increase in the β-receptor with propranolol, you still have the same rise in blood pressure.

Brodde: It was a slight, but significant decrease by 10 mm Hg.

Massry: The changes were almost not different before and after propranolol. If, then, the decrease in β-receptors is important in blood pressure in dialysis patient, why when you blocked these receptors, did the blood pressure still go up?

Brodde: No, I did not say the β-receptors are per se important for blood pressure. What I said is that these exercise experiments in dialysis patients demonstrate that acute regulation of the β-receptors is altered. In healthy volunteers a rapid and vigorous increase in sympathetic tone (for example, by dynamic exercise) leads first to an increase in β-receptors and then, later on, to a 'down-regulation', while the α-receptor is directly down-regulated under these conditions. Such a regulation does not function in hypotensive dialysis patients. They cannot completely respond to such rapid increases in the sympathetic activity.

Drueke: We and others have observed that chronic hypotension in dialysis patients may be responsive to increasing dialysate sodium concentration. Have you or others reported on β- and α-receptor numbers, and their responsiveness in such patients in whom chronic hypotension was partially corrected by increasing dialysate sodium?

Brodde: Unfortunately, not.

Heidland: I had the same question. Are there any relations between sodium state and the receptor numbers?

Brodde: Unfortunately, up to now we did not calculate it. Sodium ions, of course, can regulate affinity and/or density of α- and β-adrenoceptors, but on the other hand, changes in the physiological range seem to be not sufficient to have any regulatory influence on α- and β-receptors. That means, if the sodium concentration is changed from zero to 200, α- and β-adrenoceptor density or responsiveness is affected, but in the physiological range of 155–135 mmol/l this change seems to be too small to be detected. Another point, of course, is that sodium deprivation or sodium load effects plasma catecholamine concentrations and that might affect receptor function.

Wizemann: Is the defect of the β$_2$-receptor only true for hypotensive dialysis patients or for all?

Brodde: It is clearer to see in hypotensive patients, but we see it also in the main part of our normotensive dialysis patients.

Wizemann: Did you actually measure the adenylate cyclase, or only cyclic AMP?

Brodde: We measured in the lymphocytes only cyclic AMP because in these cells cyclase measurements are rather difficult; but we have two tools which are non-receptor-mediated and act directly at the cyclase, that is forskolin and sodium fluoride and since the effects of both drugs on the cyclase are decreased, we think it is justified to conclude, that in fact the cyclase activity is diminished.

Massry: Mr. Brodde, would you care to speculate why in uremia in dialysis these changes in the receptors occur? What are your thoughts about this?

Brodde: We think that the decreased α-adrenoceptor response in hypotensive dialysis patients is caused by the increased endogenous catecholamines, since it is well known that the catecholamines can regulate α-adrenoceptors. A long-term effect of catecholamines on α-receptors leads to a 'down-regulation'. Why the β-receptor function is decreased, at the moment we do not know. It seems to us, that this decrease is due to internalisation of the receptors. We have some evidence from, unfortunately, only one or two experiments that after pre-treatment of our hypotensive patients with prednisone, we get an exercise-induced increase in β-receptors very similar to the normotensive control. This favours the idea, that β-receptor is internalized, not destroyed. But why we don't know.

Ritz: Does PTH affect the receptor?

Massry: We are just studying. We do not have the answer.

Brodde: We did not investigate PTH effects.

Ritz: You showed age-dependent changes of the isoprenaline response in lymphocytes. Did you also examine, using forskolin, whether there are age-dependent changes of the catalytic unit or of the coupling process?

Brodde: It has been shown by Abrass and Scarpace [J. clin. Endocr. Metab. *55:* 1026, 1982] who measured directly the activity of the Ns protein in human erythrocytes, that coupling from β-receptor to the cyclase is not changed in the elderly. On the other hand forskolin stimulated cyclase was decreased, which clearly demonstrates, that in the elderly the activity of the catalytic unit of the cyclase is diminished.

Kokot: Have you some information on the activity of the renin-angiotensin system in your patients, that means in your hypotensive group?

Brodde: The plasma renin activity in dialysis patients is − despite a great variability − higher than in healthy volunteers. However, in contrast to healthy volunteers, in dialysis patients isoprenaline fails to increase PRA activity, which supports our view, that in dialysis patients β-receptor function is impaired.

Kokot: In any case your patients were not diabetic?

Brodde: No.

Ritz: There are no further questions. Thank you, Dr. Brodde, for your contribution.

Contr. Nephrol., vol. 50, pp. 36–45 (Karger, Basel 1986)

Prostanoids and Renal Failure

A Hypothetical Role for Prostanoids in Progressive Renal Disease

Michael A. Kirschenbaum

Hypertension Section, Division of Nephrology, UCLA School of Medicine, Los Angeles, Calif., USA

Since the observations of Richard Bright, nephrologists have been puzzled by the inexorable progression of diseases which affect the kidneys. Widely diverse disorders (e.g. renal disease associated with immunologic injury, diabetes mellitus, hypertension, reflux nephropathy, unilateral nephrectomy) appear to possess or share a common fundamental pathogenetic mechanism. The development of technical advances over the past two decades allowing for the precise measurement of the determinants of the rate of ultrafiltration of individual nephrons has provided data which appears to offer a new insight into the cause of progressive nephron damage. These data suggest that a reduction of nephron mass of certain renal diseases leads to, or are associated with, a circulatory adaptation in renal plasma flow in the remaining nephrons [1, 4, 5]. This adaptation may lead to a deleterious 'trade-off', the development of renal hyperfiltration, which may be a common pathogenetic event in many of these progressive disorders. Thus, despite a wide variation in the etiology of renal diseases, one single event, single nephron hyperfiltration, may be a common factor responsible for the development of glomerular sclerosis and the unrelenting destruction of countless normal nephrons.

There are several possible endogenous chemical or hormonal mediators for these hemodynamic, functional, and eventually structural changes. Of all of these endogenous substances, one group has been given considerable attention: the family of unsaturated fatty acids called the prostanoids. One or more of these lipids could function as a local regulator(s) of glomerular perfusion and thus of progressive nephron damage.

*Evidence for Dependency of Renal Perfusion, Filtration Rate, and
Hyperperfusion on Prostanoids: Studies in Intact Animals*

In general, the infusion of either PGE_2 or PGI_2 into the renal artery results in renal vasodilatation, while TxA_2 results in vasoconstriction. However, under normal circumstances, prostanoids probably enter neither the systemic nor renal circulation in concentrations comparable to those needed to have direct effects. Thus, unlike hormones produced by other endocrine organs, prostanoids rarely reach plasma concentrations which can cause systemic effects and although of interest, the infusion of prostanoids into the renal artery does not necessarily generate data analogous to the effect of endogenous synthesis and release of these lipids by renal tissue. Almost 10 years ago we began a series of studies aimed at understanding the role of these lipids in the control of renal hemodynamics. We observed that the administration of either indomethacin or meclofenamate to anesthetized dogs resulted in a 40−61% fall in total renal blood flow as well as a redistribution of regional blood flow away from inner cortical nephrons [8]. However, similar studies in conscious animals could not reproduce the significant fall in renal blood flow noted during anesthesia [7]. These and other studies led to an understanding of a primary action of renal prostanoids: the modulation of the vasoconstrictive influence of pressor hormones on the renal circulation [11]. This effect is most easily demonstrated in models in which a cyclo-oxygenase inhibitor is administered to block the formation of the prostanoids allowing the full potential effect of the vasoconstrictive influence to be 'unmasked'. The mechanism for this action may be dependent on the ability of endogenous pressors (e.g. angiotensin II, AVP, catecholamines) and α-adrenergic nerve activity to enhance prostanoid biosynthesis by their stimulatory effect on phospholipase A_2 activity. The specific arachidonic metabolites responsible for this unique biologic feedback system are not known. Either PGI_2 or PGE_2 would be logical choices since both have been shown to be produced locally in the afferent arteriole and glomerulus and both are vasodilators.

Because of the close association between glomerular perfusion and filtration rate, it is not surprising that changes in prostanoid synthesis may have a profound influence on the GFR and that these changes are more pronounced in states in which there may be prominent vasoconstrictive influence. In man, inhibition of prostanoid synthesis by administration of nonsteroidal antiinflammatory drugs (NSAIDs) results in marked depression of GFR in the setting of volume depletion or low Na diet [3] or with a

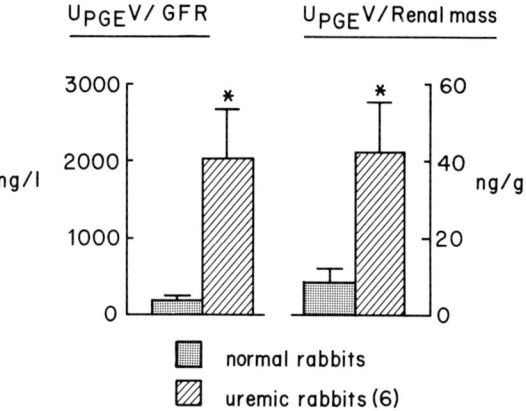

$U_{PGE}V/GFR$ $U_{PGE}V/Renal\ mass$

Fig. 1. Urinary PGE excretion levels ($U_{PGE}V$) factored by either GFR or remaining renal mass in 12 normal and 6 uremic rabbits. *p < 0.001.

variety of renal diseases such as systemic lupus erythematosus [9]. Comparable changes are seen in dogs with reduced cardiac output [12]. Similar observations have been made in this laboratory [8].

A series of studies were performed in this laboratory to assess the effect of alterations in prostanoid synthesis on GFR in rabbits with either normal renal function or after the surgical reduction of 7/8 renal mass [6]. Total renal prostanoid production was estimated by measuring 24-hour urine PGE excretion rates. In the group with normal renal function, GFR was unaltered by the administration of either indomethacin or meclofenamate despite a 53% reduction in urinary PGE excretion. However, in the uremic animals, initial GFR was 71% lower than in the normal group and basal urinary PGE excretion rates were over twice that seen in the normal group. When factored by either GFR or remaining renal mass (fig. 1), urinary PGE excretion rates were increased more than 9 × and 4 ×, respectively. Unlike the normal animals in which no change was noted, the administration of cyclo-oxygenase inhibitors resulted in a further 53% fall in GFR. Although it is clear that rates of urine PGE excretion may not accurately assess either total renal or renal cortical prostanoid production, these data are interesting not only because they suggest a fundamental role of prostanoids in the control of GFR (and presumably renal blood flow) in states of progressive renal damage but they suggest an enormous magnification in the rate of prostanoid excretion and, presumably, synthesis in these animals (fig. 1).

Table I. Effect of prostanoid inhibition (PI) of glomerular filtration rate (GFR) in normal and diabetic rats

Group	n	GFR, ml/min
Control	9	1.29 ± 0.18
Control + PI	6	1.43 ± 0.31
Diabetes	7	1.91 ± 0.23*
Diabetes + PI	7	1.30 ± 0.20

* $p < 0.02$.

Using the technique of estimating renal prostanoid production by measuring the rates of urinary PGE excretion, the effect of prostanoid inhibition on glomerular hyperperfusion was examined in another group of animals in which streptozotocin was administered in a standard manner to produce experimental diabetes mellitus. The animals were given small doses of insulin daily to maintain blood glucose levels between 200 and 250 mg/dl. Beginning with the day after the streptozotocin administration, one group was treated with a cyclo-oxygenase inhibitor in their drinking water while the other received vehicle. An additional two groups not given streptozotocin were treated in a similar manner. The results of the inulin clearance determinations for the four groups performed after 30 days are shown in table I. The GFR values in the normal and prostaglandin-inhibited normal animals were not different. However, the diabetic animals showed a marked increase in their GFR which was associated with an increase in urinary PGE excretion (not shown in the table; both $p < 0.05$ when compared to the control, nondiabetic group) and presumably, total renal prostanoid production. The administration of a cyclo-oxygenase inhibitor reduced urinary PGE excretion and GFR levels to that of the control animals. These data suggest that in diabetic rats with intact prostanoid synthesis, hyperfiltration (and presumably hyperperfusion) occurs. Inhibition of prostanoid synthesis in diabetic rats (but not in nondiabetic rats) blunted the hyperfiltration (and presumably the renal hyperperfusion) and supports an important role for prostaglandins mediating the circulatory adaptations which occur in this model.

Thus, in two experimental models demonstrating the hyperfiltration phenomenon (i.e. streptozotocin-induced diabetes mellitus and unilateral

Table II. Glomerular thromboxane (TxB$_2$) and 6-keto-PGF$_{1\alpha}$ synthesis in normal rats and in rats 1 week after streptozotocin administration

Group	TxB$_2$[1]	6-Keto-PGF$_{1\alpha}$[1]
Control	1.9 ± 0.2	1.9 ± 0.2
Diabetes	$1.2 \pm 0.1^*$	$2.5 \pm 0.2^*$

* $p < 0.001$.
[1] Results shown as pmol prostanoid/mg protein/h.

nephrectomy), experimental evidence suggests that renal prostaglandins may have an important and pivotal role in the expression of the adaptive circulatory response, the increase in glomerular perfusion.

Evidence for Prostanoid Synthesis by the Renal Microvasculature: Studies in Isolated Glomeruli

If an arachidonic acid metabolite is a tissue mediator of the hyperfiltration phenomenon, it would be imperative that it must be capable of being synthesized locally at its site of action. Although the kidney is a major producer of prostanoids, only recently have laboratories been able to demonstrate prostanoid biosynthesis by the microvasculature and isolated nephron segments. It is unlikely that prostaglandins produced in tubular segments have a significant direct influence on glomerular function. More likely, prostanoids produced by the preglomerular microvasculature (i.e. the afferent arterioles) or by the glomerular capillaries themselves could function as regulators of glomerular perfusion.

We have recently performed a series of studies aimed at evaluating whether the changes which occur in glomerular prostanoid synthesis in early streptozotocin-induced diabetes mellitus would be consistent with a role for these lipids in glomerular hyperperfusion [2]. Prostanoid production was determined by RIA in glomeruli obtained from rats after administration of either streptozotocin (60 mg/kg i.v.) or vehicle. Glomeruli were incubated for 1 h at 37°C in the presence of endogenous substrate. After 7 days, a significant increase in 6-keto-PGF$_{1\alpha}$ and a decrease in TxB$_2$ production was noted, as shown in table II. No significant differences were noted in the pro-

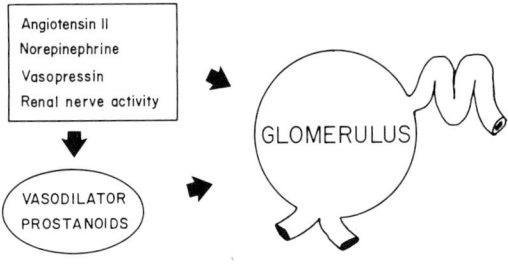

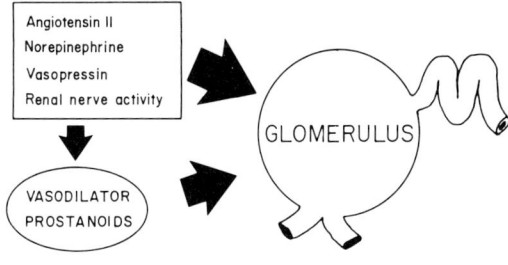

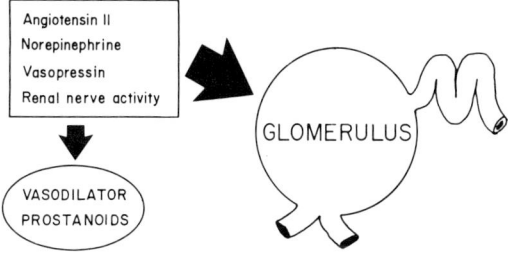

Fig. 2. Hypothetical model showing the relative importance (shown as the size of the arrows) of vasodilatory prostanoids on glomerular function in normal subjects (upper panel), subject with renal failure (middle panel), and subject with renal failure during administration of nonsteroidal antiinflammatory drugs (NSAIDs, lower panel).

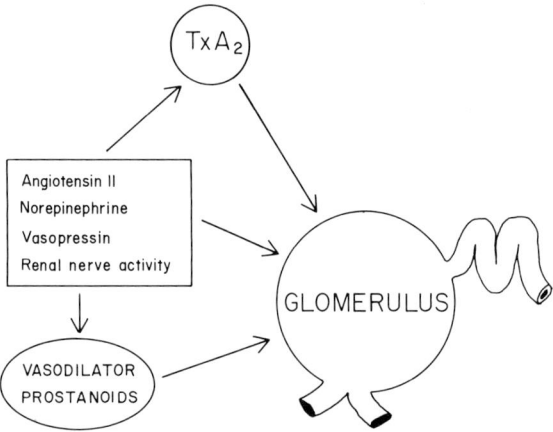

Fig. 3. Hypothetical model showing the effect of vasodilatory and vasoconstrictor prostanoids on glomerular function.

duction of either PGE_2 or $PGF_{2\alpha}$ at this time. An additional group of 6 streptozotocin-treated animals were examined at the end of 2 weeks and revealed that synthesis of both PGE_2 and $PGF_{2\alpha}$ was significantly increased. TxB_2 and 6-keto-$PGF_{1\alpha}$ synthesis was similar at 1 and 2 weeks. Similar changes have been reported by Kreisberg and Patel [10] and by Schambelan et al. [13].

Hypothetical Role of Prostanoids in the Control of Glomerular Perfusion in Progressive Renal Disease

There is evidence in man and experimental animals to indicate that there is an increase in glomerular prostanoid production associated with progressive renal damage. This increase appears to be a physiologic adaptation which serves to optimize nephron glomerular filtration in the face of a loss of nephron function. These changes are shown diagrammatically in the three panels of figure 2. Under normal conditions and, presumably, in the absence of a major influence of endogenous vasoconstrictive factors (upper panel), there appears to be only a minor role for glomerular prostanoids in the alteration glomerular perfusion. In renal failure (middle panel), however, an increase in glomerular vasodilator prostanoid production is seen in response to a series of physiologic changes (including probably an increase in the influence of vasoconstrictive factors). The net result is a hemodynamic adaptation allowing for an increase in SNGFR. This adaptation may occur at the expense of an increase in glomerular flow and pressure, and a decrease in K_f, which may be the basis, in part, of the development of progressive glomerular sclerosis and continued nephron loss. The efficiency of this circulatory adaptation may be dependent on numerous factors including those which may influence the ability of the glomerulus to produce these vasoactive lipids. This can be seen in the lower panel of figure 2 in which a decrease in glomerular function may be seen during the concomitant administration of a prostanoid inhibitor (NSAID) during renal failure. Finally, if the observations in diabetes and other renal diseases are applicable to renal failure in general, the increased production of thromboxanes by the glomerulus as it adapts to a decrease in renal function (fig. 3) may have additional deleterious long-term effects on glomerular function through a further decrease in the surface area of the glomerular capillary through its interactions with platelet aggregation.

References

1   Brenner, B.M.; Meyer, T.W.; Hostetter, T.H.: Dietary protein intake and the progressive nature of kidney disease. The role of hemodynamically mediated glomerular injury in the pathogenesis of progressive glomerular sclerosis in aging, renal ablation, and intrinsic renal disease. New Engl. J. Med. *307:* 652–659 (1982).

2   Chaudhari, A.; Kirschenbaum, M.A.: Effect of experimental diabetes mellitus (DM) on eicosanoid biosynthesis in isolated rat glomeruli. Kidney int. *25:* 326 (1984).

3   Dunn, M.J.: Renal prostaglandins: influences on excretion of sodium and water, the renin-angiotensin system, renal blood flow and hypertension; in Brenner, Stein, Hormonal function and the kidney, pp. 89–114 (Churchill Livingstone, New York 1979).

4   Hostetter, T.H.; Olson, J.L.; Rennke, H.G.; Venkatachalam, M.A.; Brenner, B.M.: Hyperfiltration in remnant nephrons. A potentially adverse response to renal ablation. Am. J. Physiol. *241:* F85–F93 (1981).

5   Hostetter, T.H.; Troy, J.L.; Brenner, B.M.: Glomerular hemodynamics in experimental diabetes. Kidney int. *19:* 410–415 (1981).

6   Kirschenbaum, M.A.; Serros, E.R.: Effect of prostaglandin inhibition on glomerular filtration rate in normal and uremic rabbits. Prostaglandins *22:* 245–254 (1981).

7   Kirschenbaum, M.A.; Stein, J.H.: The effect of inhibition of prostaglandin synthesis on urinary sodium excretion in the conscious dog. J. clin. Invest. *57:* 517–521 (1976).

8   Kirschenbaum, M.A.; White, N.; Stein, J.H.; Ferris, T.F.: Redistribution of renal cortical blood flow during prostaglandin inhibition. Am. J. Physiol. *227:* 801–805 (1974).

9   Kimberly, R.P.; Bowden, R.E.; Keiser, H.R.; Plotz, P.H.: Reduction of renal function by newer non-steroidal anti-inflammatory drugs. Am. J. Med. *64:* 804–807 (1978).

10   Kreisberg, J.I.; Patel, P.Y.: The effects of insulin, glucose and diabetes on prostaglandin

production by rat kidney glomeruli and cultured glomerular mesangial cells. Prostaglandins Leukotrienes Med. *11:* 431–442 (1983).

11   Levenson, D.J.; Simmons, C.E.; Brenner, B.M.: Arachidonic acid metabolism, prostaglandins and the kidney. Am. J. Med. *72:* 354–374 (1982).
12   Oliver, J.A.; Sciacca, R.R.; Pinto, J.: Participation of the prostaglandins in control of renal blood flow during acute reduction of cardiac output in the dog. J. clin. Invest. *67:* 229–239 (1978).
13   Schambelan, M.; Blake, S.; Nivez, M.-P.; Sraer, J.; Ardaillou, N.: Prostaglandin production is increased in glomeruli isolated from rats with streptozotocin-induced diabetes mellitus. Clin. Res. *31:* 440A (1983).

Prof. Michael A. Kirschenbaum, MD, Chief, Nephrology Section,
Veterans Administration Medical Center, Long Beach, CA 90822 (USA)

Discussion

Ritz: Thank you, Dr. Kirschenbaum, for giving us a panoramic view over a large area. May I start by asking you. You mentioned the paper of Cinotti's group in *New England Journal of Medicine* in which it was stated that sulindac was selective in having extrarenal effects without any intrarenal action. To the best of my knowledge, this has been challenged by several groups and there has been a paper in one of the last issues of *Kidney International* where the clearest renal effects of sulindac could be demonstrated. Since it is of prime importance to the nephrologist, whether there exists a unique cyclo-oxygenase inhibitor which is not acting on the kidney, could you briefly comment and elaborate on this point?

Kirschenbaum: The drug sulindac appears, at least in vitro, to lack the ability to inhibit cyclo-oxygenase in renal cells, either because it cannot be internalized into the cells to inhibit cyclo-oxygenase or, if it does get into the cells, does not have the ability to be metabolised to the active form of the drug which then goes on and inhibits prostanoid production. I read with great interest the *Kidney International* article that you are referring to because it almost showed the opposite of what previous data would suggest. I understand that in some people's hands, and primarily in the hands of the Italians, sulindac, when administered to healthy volunteers, leads to no decrease in urinary prostaglandin secretion, yet it does lead to a decrease in platelet thromboxane production. The clinical observations which have been made in our country and here in Europe are that the drug sulindac is associated with a lower incidence of renal toxicity. So, although we have all seen cases of these haemodynamically-mediated acute renal failure episodes associated with sulindac, it appears, concerning the number of doses that are being used in our country, that we should be seeing more if the drug was indeed equally as nephrotoxic as the other agents. The fact that we do not is very suggestive that there is some basic metabolic difference between the drug and other non-steroidals. I do not know how to explain the data in *Kidney International* which seemed to show that ibuprofen had less inhibitory activity on renal cyclo-oxygenase than sulindac did. I do not have an explanation for those data.

Brodde: I would like to come back to the question that Dr. Ritz asked earlier in the morning. If norepinephrine really plays an important role in renal failure, one should be able to improve renal function by α-blockers. Is there any evidence? Secondly, how does presynaptic control of catecholamine release fit in your scheme? I think angiotensin II is fascilitating noradenalin release, while I think prostaglandins are inhibitory on noradrenalin release. How does this fit?

Kirschenbaum: It is very confusing. You may be better able to answer these questions than I am. The interactions between the α_2- and α_1-receptor and the β-receptor and the prostaglandins is a very complicated one, because it is difficult to dissect apart which of the actions that one is looking at, i.e. whether one is looking at a primary action or the result in a secondary one from the receptor interactions, and then the response to them. It is clear that at least α-agonists and many β-agonists do result in increased prostanoid production in β-selective tissues that have been examined. How this fits into an overall scheme or how particular organs work is unknown. Thus, I am not going to be able to answer that question because I am not sure how these interactions are really related. There is evidence in the presynaptic membrane that norepinephrine can regulate its own secretion by release of prostaglandins. However, many individuals who are in this field have told me that these data are very soft and it may not actually be the case. On the postsynaptic membrane there is again suggestive evidence, at least in the α-system, that binding to the α-receptor may again cause some interactions with the production of naturally-occurring ionophoretic substances that may then increase calcium entry into cells and this may increase prostaglandin activity. There are so many small pieces of data that cannot be put together into a cohesive model, mainly because everyone is looking at his own field and not looking at all of the potential interactions. That is not an answer to your question, but I do not think there is any answer right now.

Gross: You showed us data of increased prostaglandin production in renal tissue after glycerol and mercury chloride, implying a physiological role of these substances. Can we be sure that these elevated levels of prostaglandins were not due to non-specific membrane damage activating phospholipases by superoxide damage, etc.?

Kirschenbaum: In some of the models of renal failure, for example those with nephrotoxic heavy metals, anatomical damage of the nephrons occurs. One can question whether the changes in prostanoid activity could be due to non-specific effects that you suggest. However, in other models, for instance, with streptozotocin-induced diabetes mellitus, with ureteral obstruction and early in heavy metal administration (within the first few hours), one might see minimal lesions predating any structural change. So it still is very possible that some of the changes seen may be non-specific, but even if they are non-specific the net effect is still the same and the physiological importance of the production of prostaglandins, even if they just happen to be occurring non-specifically, can be seen just with the use of non-steroidals because when one blocks the presence of those prostaglandins one sees dramatic worsening of renal function. So even if this is just an accidental effect of some membrane activation of phospholipases, the net effect is the body utilises those changes. To prevent the body from having access to these vasodilated prostaglandins leads to an enormous deterioration of renal function.

Rosenblatt: I was intrigued by the very nice data showing that intracellular calcium stimulates prostaglandins synthesis and I was wondering, since that raises the whole question of the phosphatidylinositol pathway, if you have had a chance to look at this system?

Kirschenbaum: Not as yet. The data I have shown are still warm. The laboratory has now finished a series of studies using calcium channel blockers to complement the ionophore studies and we intend to investigate the entire phospholipase system, piece by piece, and then look into potential biological stimulators, such as PTH and other naturally-occurring ionophores.

Contr. Nephrol., vol. 50, pp. 46–53 (Karger, Basel 1986)

Secretion and Catabolism of Antidiuretic Hormone in Renal Failure

Raymond Ardaillou, Wojciech Pruszczynski, Mustapha Benmansour

Inserm U-64, Hôpital Tenon, Paris, France

Renal failure can be considered as a model allowing analysis of the role of the kidney in the excretion and the catabolism of any hormone in humans. We also felt in the particular case of antidiuretic hormone (ADH) that the study of uremic patients treated by hemofiltration was a useful tool to investigate the ADH response to the volume stimulus and the role of the renin-angiotensin system in the control of this response.

Catabolism of Antidiuretic Hormone in Advanced Chronic Renal Failure

Basal concentration of ADH is higher in patients with advanced renal failure on maintenance hemodialysis than in normal subjects [1–3]. This occurs in spite of the loss of ADH across the dialysis membrane during hemodialysis sessions, as shown by Shimamoto et al. [4]. The main reason for this persistent elevation of plasma ADH is the decrease in the catabolism of the hormone due to the loss of functional renal tissue. The role of the kidneys in the breakdown of all polypeptide hormones is well established [5]. The metabolic clearance rate of immunoreactive arginine vasopressin (AVP) and [^{131}I]iodovasopressin was diminished in binephrectomized dogs [6] and it has been shown that the isolated perfused rat kidney destroyed ADH taken up from the medium [7]. We have demonstrated the role of the kidney in ADH catabolism in humans by measuring the metabolic clearance rate (MCR) of ADH after injection of synthetic AVP in 10 uremic patients and 16 healthy subjects [2]. The disappearance curves of ADH in the plasma were drawn after determination of ADH by radioimmunoassay.

The curves could be resolved as the sums of two exponentials. The volume of distribution and the MCR of the hormone were obtained from the calculated parameters (table I). For a period starting at the 10th min, the mean plasma concentration of AVP was always higher in the uremic patients than in the control subjects. The mean half-life of the low-slope exponential was 180 ± 17 and 78 ± 5 min in the uremic and the healthy subjects, respectively. In parallel, the MCR was much smaller in uremic (102 ± 15 ml/min/ m^2) than in healthy (287 ± 42 ml/min/m^2) subjects. We also observed that the volume of rapid distribution of AVP was greater in uremic (293 ± 38 ml/kg) than in normal (219 ± 19 ml/kg) subjects. The result observed in the control group was close to the theoretical value of the extracellular fluid volume. The increase in the uremic patients can be attributed to the usual expansion of the extracellular compartment just prior to a dialysis session. The results for MCR observed in both groups show that renal catabolism accounts for about two-thirds of the total metabolic clearance rate of ADH since the value for uremic patients represents about one-third of that obtained in normal subjects. The nonrenal part of MCR depends probably of the liver. Studies on the extraction of ADH by the isolated perfused kidney [7] and liver [8] of the rat lead to similar estimations of the role of both organs: 35 and 65% for the liver and the two kidneys, respectively. The measurement of MCR of ADH in uremic and normal subjects provides no direct information on the handling of this hormone by the kidney. One can only assume that ADH is taken up by the renal cells both from urine on their luminal side and from plasma on their basolateral pole since the renal contribution to the total MCR of ADH (i.e. the difference between the total MCR of normal subjects and uremics) equals about 330 ml/min. This

Table I. Metabolic clearance rate of ADH in humans (mean ± SEM): results reproduced from Benmansour et al. [2]

	Basal plasma ADH pg/ml	Metabolic clearance rate ml/min/m²	Volume of diffusion ml/kg	Half-life 2nd exponential min
Healthy subjects (n = 16)	3.4 ± 0.3	287 ± 42	219 ± 19	77.6 ± 4.6
Patients with advanced renal failure (n = 10)	6.3 ± 1.1	102 ± 15	293 ± 38	179.8 ± 16.8

value is clearly higher than the normal glomerular filtration rate in man, which suggests that uptake of ADH from the plasma may well occur.

Renal Excretion of Antidiuretic Hormone in Moderate Chronic Renal Failure and in Acute Renal Failure at the Polyuric Phase

In order to better analyze the role of tubular absorption in the renal handling of ADH, we have measured the urinary clearance of this hormone expressed both as absolute value and as fraction of glomerular filtration rate in healthy subjects; patients with moderate renal failure were who considered as a model of nephron reduction without glomerulotubular disequilibrium and patients with acute renal failure at the polyuric phase who were considered as a model of glomerulotubular disequilibrium [9]. The results are indicated in table II. In patients with mild chronic renal failure, urinary excretion rate and urinary clearance of ADH were lower than in normal subjects whereas plasma ADH and the fractional clearance of ADH were unchanged. These results are the consequence of the reduction in the number of functional nephrons, each of them exhibiting a normal reabsorptive capacity which can be estimated as representing a small fraction (6–10%) of the amount of filtered ADH. In patients with acute renal fail-

Table II. Renal excretion of antidiuretic hormone (mean ± SEM): results reproduced from Pruszczynski et al. [9]

	Plasma concentration of ADH pg/ml	Urinary excretion rate of ADH pg/min	Urinary clearance of ADH ml/min	Fractional clearance of ADH %	Urinary clearance of creatinine ml/min
Healthy subjects (n = 10)	3.3 ± 0.36	25.2 ± 5.5	75 ± 1.2	6.4 ± 1.0	116 ± 6.8
Patients with moderate chronic renal failure (n = 18)	2.8 ± 0.19	9.4 ± 2.0	3.4 ± 0.6	10.0 ± 2.9	45 ± 5.9
Patients with acute renal failure (n = 7)	4.6 ± 0.47	52.8 ± 15.8	9.5 ± 2.7	24.9 ± 4.4	36.1 ± 6.4

ure, urinary excretion rate and fractional clearance were higher than in healthy subjects and patients with chronic renal failure. This may be interpreted as the consequence of the glomerulotubular disequilibrium observed at the stage of polyuria in acute renal failure. At this stage, glomerular filtration operates again whereas there is still no complete recovery of the tubular function. Since ADH which is a small peptide is entirely filtered but is recovered only in small amount in the final urine, except in patients with acute renal failure, it must be concluded that the hormone is extracted from the tubular urine. As for other peptides [5], it is likely that luminal removal is achieved through absorption across the luminal membrane and intracellular catabolism by degrading enzymes.

Antidiuretic Hormone Response to the Volume Stimulus in Uremic Patients Treated by Hemofiltration

The secretion of ADH is regulated essentially by changes in plasma osmolality and plasma volume. Robertson et al. [10] have analyzed the functional characteristics of the osmoreceptors controlling ADH secretion by examining the relationship between plasma ADH and plasma osmolality over a wide range of values. The threshold and the sensitivity of the system were estimated from the abscissa intercept and the slope of the regression line obtained by plotting plasma ADH versus plasma osmolality. The role of the volume stimulus per se had not been clearly defined in man because it was more difficult than for the osmotic stimulus to quantify the relationship between the decrease step by step in volume at constant osmolality and the resulting increase of plasma ADH. We thought that hemofiltration which produces a decrease in plasma volume at a rapid rate without any change in osmolality was a means to overcome this difficulty. In a first study of 3 uremic patients treated by hemofiltration [1], we observed that there was a linear relationship between the increase in plasma ADH and the quantity of ultrafiltrate. For every 100 ml of water lost, there was a rise of 0.55 pg/ml of plasma ADH. The minimum fluid loss required to result in an increase of plasma ADH was 320 ml. In a subsequent study of 4 patients [11] a neighboring value was observed for the slope, 0.36 pg/ml for each 100 ml water lost but no threshold could be demonstrated. In both studies 0.16 M NaCl was administered intravenously after marked depletion had been obtained. This produced an immediate and rapid fall in plasma ADH followed by a slower decrease.

Table III. ADH response to the volume stimulus in uremic patients

	3 uremic patients [Caillen et al., 1]	4 uremic patients [Benmansour et al., 11]	4 uremic patients pretreated with captopril [Benmansour et al., 11]	3 binephrectomized uremic patients [Caillen et al., 1]
Slope, pg/ml/100 ml	0.55	0.36	0.14	0.24
Threshold, ml	320	negligible	negligible	820

These parameters were calculated from the regression lines between the increase in ADH concentration expressed as pg/ml and the quantity of ultrafiltrate expressed as 100 ml.

The model of the uremic patient also allowed us to demonstrate the essential role of angiotensin II as a mediator of the volume stimulus in man. Two experimental protocols were used [1, 11]. On the one hand, the effect of an identical volume depletion was considerably smaller in binephrectomized uremic patients than in those with kidneys remaining. The threshold was greater and the slope was flatter. On the other hand, pretreatment of the patients with captopril, an inhibitor of converting enzyme, followed by ultrafiltration 60 min later resulted in a marked diminution of the slope when compared to that obtained in the same patients employing the same protocol but without prior administration of captopril. These results are summarized in table III. In contrast, angiotensin II does not seem to be a mediator of the osmotic stimulus since prior administration of captopril to patients with chronic renal failure did not change the response to administration of hypertonic saline [11]. We are quite aware that it is not certain that results obtained in patients with terminal renal failure can be totally transposed to healthy subjects. In particular, cardiac autonomic dysfunction appears to occur in chronic uremia and is probably related to the generalized neuropathy observed in this state [12]. We have recently shown that plasma ADH response to administration of furosemide and passage from lying to standing was suppressed in diabetic patients with cardiac autonomic dysfunction whereas it was maintained in intact diabetic patients [13]. Thus, the effect of this disorder would be to diminish ADH response to hypovolemia in the uremic patients studied.

References

1   Caillens, H.; Pruszczynski, W.; Meyrier, A.; Ang, K.S.; Rousselet, F.; Ardaillou, R.: Relationship between change in volemia at constant osmolality and plasma antidiuretic hormone. Mineral Electrolyte Metab. 4: 161−171 (1980).

2   Benmansour, M.; Rainfray, M.; Paillard, F.; Ardaillou, R.: Metabolic clearance rate of immunoreactive vasopressin in man. Eur. J. clin. Invest. 12: 475−480 (1982).

3   Shimamoto, K.; Watarai, I.; Miyahara, M.: A study of plasma vasopressin in patients undergoing chronic hemodialysis. J. clin. Endocr. Metab. 45: 714−720 (1977).

4   Shimamoto, K.; Ando, T.; Nakao, T.; Watarai, I.; Miyahara, M.: Permeability of antidiuretic hormone and other hormones through the dialysis membrane in patients undergoing chronic hemodialysis. J. clin. Endocr. Metab. 45: 818−820 (1977).

5   Ardaillou, R.; Paillard, F.: Métabolisme rénal des hormones polypeptidiques; in Grunfeld, Actualités néphrologiques de l'Hôpital Necker, pp. 186−205 (Flammarion, Paris 1979).

6   Shade, R.E.; Share, L.: Metabolic clearance of immunoreactive vasopressin and immunoreactive ^{131}I iodovasopressin in the hypophysectomized dog. Endocrinology 99: 1199−1206 (1976).

7   Rabkin, R.; Share, L.; Payne, P.A.; Young, J.; Crofton, J.: The handling of immunoreactive vasopressin by the isolated perfused rat kidney. J. clin. Invest. 63: 6−13 (1979).

8   Rabkin, R.; Ghazelah, S.; Share, L.; Crofton, J.; Unterhalter, S.A.: Removal of immunoreactive arginine-vasopressin by the perfused rat liver. Endocrinology 106: 930−934 (1979).

9   Pruszczynski, W.; Caillens, H.; Drieu, L.; Moulonguet-Doleris, L.; Ardaillou, R.: Renal excretion of antidiuretic hormone in healthy subjects and patients with renal failure. Clin. Sci. 67: 307−312 (1984).

10 Robertson, G.L.; Mahr, E.A.; Athar, S.; Sinha, T.: Development and clinical application of a new method for the radioimmunoassay of arginine vasopressin in human plasma. J. clin. Invest. 52: 2340−2352 (1972).

11 Benmansour, M.; Caillens, H.; Ardaillou, R.: Effets de l'inhibition de la synthèse de l'angiotensine II sur la sécrétion de l'hormone antidiurétique chez l'urémique en réponse aux stimuli osmotique et volumique. Nephrologie 1: 109−112 (1980).

12 Ewing, D.J.; Winney, R.: Autonomic function in patients with chronic renal failure on intermittent hemodialysis. Nephron 15: 424−429 (1975).

13 Grimaldi, A.; Pruszczynski, W.; Thervet, F.; Ardaillou, R.: Antidiuretic hormone response to volume depletion in diabetic patients with cardiac autonomic dysfunction. Clin. Sci. 68: 545−552 (1985).

Prof. Dr Raymond Ardaillou, Hôpital Tenon, 4, rue de la Chine,
F−75970 Paris Cedex 20 (France)

Discussion

Gurland: Thank you, Dr. Ardaillou.

Kurtz: Can it be concluded from your data that the response of ADH release upon water loss is mediated by angiotensin II?

Ardaillou: I think so. In order to get a full response you need an intact renin-angiotensin system as demonstrated by many experiments. In particular, it has been shown by the group of Vincent (Bordeaux, France) that direct exposure to angiotensin II of the AVP-secreting neurones in the monkey produced an increase in the electrical activity of these cells. There was in parallel a stimulation of AVP release.

Drueke: You showed that the increase in plasma ADH during volume depletion in haemodialysis patients is rather progressive. However, during volume restitution, the decrease in ADH is very rapid – let us say within 500:1,000 ml? Why is there that difference in the evolution of the curve?

Ardaillou: I cannot explain this difference. What I can do only is to quote other results demonstrating a similar difference between the time-courses of AVP release and of its stimulus. It is well known that water load results in a decrease in plasma osmolality and plasma AVP; but there is an immediate fall in plasma AVP at a time when plasma osmolality has not been yet modified. This suggests the presence of osmoreceptors in the mouth or in the stomach. In this study, as in ours, there is no simultaneity between the changes in plasma ADH and in the plasma stimulus of ADH release.

Ritz: I was struck by the magnitude of non-filtration clearance in the kidney in the range of 100–150 ml/min. Do you have any explanation or any concepts about the mechanism? Is this receptor-mediated? How is the peritubular uptake achieved in the kidney?

Ardaillou: Similar results have been observed for other polypeptide hormones such as insulin, calcitonin and parathyroid hormone. It has been shown that for most polypeptide hormones the bulk of the filtered hormone is reabsorbed in the proximal tubule. Only a small part of the hormone is excreted. I think, for that reason, that assay of urinary ADH cannot be considered as a good method for estimation of AVP secretion.

Ritz: That was not the question. The question was why is peritubular uptake so high for parathyroid hormone or for insulin? It is 30% of the filtered, but here it is almost exceeding.

Ardaillou: The renal handling of AVP is the same as that of insulin and calcitonin. Filtered AVP is taken up in the proximal tubule and destroyed within the proximal tubular cells. This is not a real reabsorption since AVP does not pass in the plasma.

Ritz: The total clearance exceeds the filtered clearance by 100%.

Ardaillou: In addition to the urinary uptake of AVP there is also an extraction of the hormone from the peritubular capillaries.

Ritz: Why does this occur? Is there a receptor?

Ardaillou: I don't think that this process is receptor-mediated since it is common to many polypeptide hormones. The renal handling of AVP has been, in particular, studied by Prof. Rabkin (San Francisco, Calif.). He demonstrated that there was a peritubular uptake of the hormone. Furthermore, AVP is entirely filtered through the glomerular capillaries and crosses haemodialysis membranes freely.

Massry: When you talk about the renal handling of ADH and you said 'reabsorbed', you do not mean it has been reabsorbed in the classical way, that it is reabsorbed into the blood?

Ardaillou: No, AVP is not reabsorbed into the plasma. It is taken up by the tubular cells and destroyed within these cells.

Massry: How do you know it is taken up by the proximal tubule?

Ardaillou: I have no personal data to affirm that AVP is taken up in the proximal tubule. Such a process has been demonstrated for other polypeptide hormones and confirmed for AVP by electron microscopy studies.

Massry: Other hormones, such as PTH, calcitonin or insulin have receptors in the proximal tubule which would facilitate their uptake.

Ardaillou: Are you sure that these binding sites are real receptors? This is not only a semantic problem. There is only one conclusive means to identify receptors. You need a family of agonists or antagonists and you have to show that the responses to these different products follow the same order in the binding and the biological studies. Such a demonstration is still lacking for the binding sites of calcitonin and insulin on the luminal membranes of the proximal cells. Thus, we can conclude presently that these binding sites exist but we cannot conclude they are real receptors.

Kokot: ADH secretion is dependent upon the changes of the effective blood volume. Have you some data about the relationship between ADH plasma levels and plasma volume changes? This is the first question. Plasma volume – how did it change during water repletion?

Ardaillou: In this study we did not measure the absolute value of plasma volume. We measured only the amount of plasma water loss, but not the absolute value.

Kokot: How did you know that it was plasma water and not extravascular water, because this is important for the interpretation of your results? This is why I will comment on Dr. Drueke's question. If you are infusing saline in the repletion period, you observe an abrupt decrease of ADH. In my opinion, it was a repletion effect of the effective plasma volume and this is decisive for ADH secretion. During the second period of water repletion, the administered water was leaving the vascular bed and going into the extravascular space which is unimportant for ADH secretion. In this way you can interpret the data you have presented here.

Ardaillou: We did not measure, in this study, the distribution of water between the extracellular and plasma volumes. It is not easy to estimate precisely acute changes in water distribution. Thus, I agree with you. Perhaps when we injected saline, the immediate increase in plasma volume accounted, at least in part, for the rapid drop in plasma AVP.

Ritz: May I come back to the haemofiltration experiment with and without captopril? You show beautifully that the major part of the response is mediated by circulating angiotensin II but there is still residual volume-mediated ADH modulation, and conventional wisdom holds that this is mediated by the baroreceptor. There has been some recent work on ADH in the direction with the natriuretic factor. Would you care to go out on a limb and tell what is the current state of the art in this field?

Ardaillou: I have no personal data on the atrial natriuretic factor (ANF). I have only personal data on the role of atrial receptors in AVP response to volume depletion. We have performed two studies in heart-transplant recipients and in diabetic patients with autonomic dysfunction. In both groups of patients, functional vagal pathways between the atriums and the central nervous system are lacking. Volume depletion was induced by a change from the lying to the standing position and by the simultaneous intravenous administration of 40 mg furosemide. In healthy control subjects, this protocol results in the stimulation of plasma AVP whereas plasma AVP is unchanged in the denervated patients. These experiments show that in humans, similarly to what ist found in dogs, cardiac receptors and cardiac innervation play a role in the ADH response to the volume stimulus.

Contr. Nephrol., vol. 50, pp. 54—63 (Karger, Basel 1986)

Vasopressin and Hyponatremia in Renal Insufficiency[1]

Peter Gross[a], Wolfgang Rascher[b]

Departments of [a]Medicine, and [b]Pediatrics, University of Heidelberg, Heidelberg, FRG

Arginine vasopressin, the human antidiuretic hormone, is a nonapeptide of defined structure and function. Since the classical description of a sensitive radioimmunoassay [1], there has been a rapid progression of vasopressin research. As a result, hydro-osmotic and vasoactive properties have been characterized in some detail. Recently, the synthesis both of effective antagonists [2] as well as of specific agonists has stimulated additional scientific curiosity. In contrast, there are only few reports concerning the role of vasopressin in renal insufficiency, even though current knowledge would suggest the potential relevance of such an association. In the following, we shall consider these issues; first of all, we outline some methodological peculiarities related to renal failure.

Metabolism of Vasopressin

Observations in normal volunteers and laboratory animals suggest that renal failure can be expected to alter the normal metabolism of vasopressin in several ways. Thus, the kidney and the liver were shown to be the two major organs that extract vasopressin from the circulation. Their extraction ratios accounted for 25 and 12%, respectively [3, 4]. As a consequence of negligible albumin binding [4], renal extraction of vasopressin is equally attributable to glomerular filtration with excretion of intact hormone as well as to peritubular uptake [5]. It has been proposed that receptor binding mediates the latter process [6], followed by the appearance of metabolic

[1] Supported by the Deutsche Forschungsgemeinschaft Gr 605/3−1 and Ra 321/1−1.

breakdown products in the circulation [5, 7]. Such hydrolyzed fragments of vasopressin may bind to the antibodies used by individual radioimmunoassays [7]. As a corollary to renal extraction, reduced renal function — as simulated by decreased renal blood flow [8] — not surprisingly is associated with a proportionately reduced renal extraction rate of vasopressin. Comparable to the known high glomerular filtration of vasopressin it has also been shown to possess considerable ultrafilterability through artifical membranes, including dialyzer membranes [9, 10]. Taken together, plasma vasopressin measurements should be interpreted cautiously in patients with renal failure, unless the radioimmunoassay used has been demonstrated to specifically exclude vasopressin fragments and unless the clearance of vasopressin across dialyzer membranes is considered in such patients.

Osmotic Control of Vasopressin

Vasopressin secretion may be modified by a number of stimuli such as drugs [11, 12], pain, nausea [13], smoking [14], hypovolemia and hypotension [15], but physiologically it is primarily under the control of plasma osmolality [15] in a highly sensitive fashion. This observation in experimental animals [15, 16] has now been tested in humans, too [17]. In this work, infusions of hypertonic saline or mannitol prompted parallel increases in plasma osmolality and vasopressin [17]. The slope describing this relationship was reproducible for each individual and defined the osmotic threshold for vasopressin secretion in healthy volunteers at approximately 280 mosm/kg H_2O [1]. It is therefore necessary to interpret plasma vasopressin concentrations in relation to the simultaneously measured osmolality [1]. In the same study, however [17, 18] infusions of hypertonic glucose or hypertonic urea were found to have a different effect: although increasing measured plasma osmolality in a similar way as hypertonic saline, these substances actually suppressed vasopressin. Thus, increases in plasma glucose or urea concentrations are ineffective vasopressin stimuli. As a consequence, effective plasma osmolality must be calculated and plotted against vasopressin concentration in such patients.

Unfortunately, several uncertainties exist as to the specifics of this transformation in uremia: (a) because of the biphasic response of vasopressin to hypertonic urea [17] it is unknown how the reduction of urea concentration during dialysis may influence vasopressin secretion, e.g. it might be a stimulus; (b) similarly, recent work in hyperglycemia has demonstrated an additional role of the simultaneous insulin status in determining the osmotic effectiveness of glucose; thus, in the presence of insulin hyperglycemia had little osmotic effect, whereas in the absence of insulin, it was fully effective; glucose intolerance is frequently observed in uremic patients, however their insulin status is usually unknown; (c) the osmolal gap has been observed to be significantly increased in chronic renal failure by approximately 8 mosm/kg H_2O [19]; the significance of these 'osmoles of uremia' for effective plasma osmolality is currently unknown; (d) it has not been studied in man whether the advent of the uremic state per se might affect slope or intercept of the vasopressin/osmolality relationship in a given individual; experimental observations however would suggest such an alteration [32]; (e) finally, the dialysis procedure itself causes many variations in the stimulation of vasopressin (vide infra), one of which appears to be related to changing plasma ionized calcium concentrations [26].

In summary, although an obligatory requirement in the interpretation of vasopressin concentrations in normal physiology, the vasopressin/osmolality relationship is incompletely understood in uremia; this adds further uncertainties to the interpretation of vasopressin in renal failure. The following discussion should be seen on the background of these problems.

Hyponatremia in Chronic Renal Failure

Several recent studies in patients [20–22], including our own work [23] have presented evidence that clinical hyponatremia is a disorder of water metabolism; in it failure to suppress antidiuretic hormone is regularly observed and appears to be crucial to the pathogenesis. Such a sequence of events would not necessarily be expected in hyponatremia of renal insufficiency, in which ADH sensitivity of the collecting duct is progressively impaired [24], while maximal free water clearance diminishes in proportion to the reduced glomerular filtration rate [25]. Thus, the amount of hypotonic fluid intake that cannot be excreted in the urine may then become the primary determinant of water balance and plasma sodium concentration [25]. Nevertheless, several authors have observed elevated basal plasma vasopressin concentrations in uremic subjects before dialysis [26–30].

To investigate this discrepancy we therefore studied hyponatremia and vasopressin in uremic subjects in our hospital. Amongst 82 consecutive, unselected hyponatremic patients on the ward, there were 10 who suffered from advanced renal insufficiency; they were not yet treated by dialysis. All of these patients showed significant hyponatremia and hypo-osmolality (effective plasma osmolality). All patients received high-dose loop diuretics and the beginning of the hyponatremia could be traced back to the start of these treatments. Their average daily intake of hypotonic fluid was in excess of 3.5 liters, while the patients had approximately 200 mM/day of sodium in their urines, probably causing a negative sodium balance. In the majority of these patients (7/10), we failed to detect evidence of plasma volume contraction; thus, their weights had been constant, their internal jugular veins were seen to be appropriately filled and there were no pathological orthostatic circulatory changes. However, plasma vasopressin concentrations were not suppressed. None of these patients had received any medications relevant to the secretion of vasopressin, nor were they under the influence of pain, nausea, or cigarette smoking when plasma samples for vasopressin determination were obtained under standard conditions. We interpret our findings as follows: The hyponatremia is most probably explained by a combination of three factors: (a) negative sodium balance from diuretics; (b) severely restricted free water clearance because of renal insufficiency, and (c) inadequate hypotonic fluid intake. The observed concentrations of plasma vasopressin were not explained by known physiologic stimuli; therefore, it cannot be excluded that metabolic breakdown products made a significant contribution to our measurements.

Role of Vasopressin during Dialysis

Whereas the hydro-osmotic effect of vasopressin is rendered unimportant by the near-complete failure of renal function, a role of vasopressin in cardiovascular homeostasis during dialysis is more than likely. Such has been pointed out by several authors [27, 29–31]. They have stressed the importance of dialysis-related volume changes in the secretion of vasopressin.

In general, there are several lines of evidence for a vascular role of vasopressin in selected circumstances: thus, exposure to vasopressin of surgically isolated arteries from the dog is followed by significant vasoconstriction [33], possibly related to specific binding sites for vasopressin on vascular smooth muscle cells [34]. A physiologic role of such a mechanism is further suggested in several studies. Dunn et al. [15] have found large diminutions of intravascular volume to cause an exponential rise of plasma vasopressin concentration, which surpassed the effect of a large osmotic stimulus. Recent studies of states of compromised intravascular volume

[35−37], utilizing vascular antagonists to vasopressin, have indeed shown a role of vasopressin in the maintainance of blood pressure. In these studies, the observed vasopressin response to vascular compromise depends, however, largely on an intact cardiac innervation [37, 38] and is an integral part of a more general stimulation of several vasoactive hormones [36]. Impairment to one of these hormones (renin, catecholamines, vasopressin) increases the sensitivity of the remaining hormonal systems to volume changes. This has also been demonstrated in humans [39]. Such a finding may be of relevance to renal failure, since renin secretion is often impaired as a consequence of reduced renal mass, while, simultaneously, the vascular response to noradrenaline may be diminished [40]. Additional evidence in humans further shows that hypotension independent of hypovolemia may also serve as a potent stimulus of vasopressin secretion [1, 41, 42]. Thus, in the study by Robertson et al. [1], a reduction of mean arterial blood pressure to 60 mm Hg reversibly increased plasma vasopressin to approximately 30 pg/ml. Taken together, there is a large body of indirect evidence indicating a potential role of vasopressin in the changes of volume and blood pressure inherent to dialysis.

A number of studies [26−32] have addressed this issue. Whereas plasma vasopressin concentrations were found unchanged, lowered [28, 29] or equivocal [30] in studies without deliberate attempts to control ultrafiltration, they were significantly increased when studies focused on this parameter [27, 31]. Nonetheless, the observed effects were small. However, it should be pointed out that the potentially most powerful stimulus to vasopressin secretion − dialysis hypotension − was not evaluated in these studies. Such hypotension is not uncommon [43, 44]; in some cases, autonomic insufficiency may contribute to hypotensive episodes during dialysis [43, 44]. In autonomic insufficiency, plasma vasopressin concentration may fail to rise during hypotension [45]. Therefore, the response of vasopressin secretion during hypotension may be an interesting index for the function of the baroregulatory reflex arc in uremia [45].

Miscellaneous

Several aspects pertaining to vasopressin function in general have met with little interest; thus, their role in uremia is unexplored. It is not known whether vasopressin may participate in any dysregulation of glomerular hemodynamics in renal insufficiency causing progressive loss of glomerular function [46]. Likewise, any potential role of vasopressin in uremic alterations of sweat production [47, 48] and gastrointestinal water absorption has not been clarified. Finally, the relationship of vasopressin to thirst [49] – though a tempting proposal – remains to be tested in uremia in the future.

References

1   Robertson, G.L.; Mahr, E.A.; Athar, S.; Sinha, T.: Development and clinical application of a new method for the radioimmunoassay of arginine vasopressin in human plasma. J. clin. Invest. *52:* 2340 (1973).
2   Manning, M.; Sawyer, W.H.: Antagonists of vasopressor and antidiuretic responses to arginine vasopressin. Ann. intern. Med. *96:* 520 (1982).
3   Lauson, H.D.; Bocanegra, M.; Beuzeville, C.F.: Hepatic and renal clearance of vasopressin from plasma of dogs. Am. J. Physiol. *209:* 199 (1965).
4   Baumann, G.; Dingman, J.F.: Distribution, blood transport and degradation of antidiuretic hormone in man. J. clin. Invest. *57:* 1109 (1976).
5   Rabkin, R.; Share, L.; Payne, P.A.; Young, J.; Crofton, J.: The handling of immunoreactive vasopressin by the isolated perfused rat kidney. J. clin. Invest. *63:* 6 (1979).
6   Weitzman, R.E.; Fisher, D.A.: Arginine vasopressin metabolism in dogs. I. Evidence for a receptor-mediated mechanism. Am. J. Physiol. *235:* E591 (1978).
7   Lindheimer, M.D.; Reinharz, A.; Grandchamp, A.; Vallotton, M.B.: Fate of vasopressin perfused into nephrons of Wistar and Brattleboro rats. Clin. Sci. *58:* 139 (1980).
8   Shade, R.E.; Share, L.: Renal vasopressin clearance with reductions in renal blood flow in the dog. Am. J. Physiol. *232:* F341 (1977).
9   Bocanegra, M.; Lauson, H.D.: Ultrafiltrability of endogenous antidiuretic hormone from plasma of dogs. Am. J. Physiol. *200:* 486 (1981).
10 Shimamoto, K.; Ando, T.; Nakao, T.; Watarai, I.; Miyahara, M.: Permeability of antidiuretic hormone and other hormones through the dialysis membrane in patients undergoing chronic hemodialysis. J. clin. Endocr. Metab. *45:* 818 (1977).
11 Suskind, R.M.; Brusilow, S.W.; Zehr, J.: Syndrome of inappropriate secretion of antidiuretic hormone produced by vincristine toxicity with bioassay of ADH levels. J. Pediat. *81:* 90 (1972).
12 Kimura, T.; Matsui, K.; Sato, T.; Yoshinaga, K.: Mechanism of carbamazepine-induced antidiuresis: evidence for release of antidiuretic hormone and impaired excretion of a water load. J. clin. Endocr. Metab. *38:* 356 (1974).
13 Rowe, J.W.; Shelton, R.C.; Helderman, J.H.; Vestal, R.E.; Robertson, G.L.: Influence of the emetic reflex on vasopressin release in man. Kidney int. *16:* 729 (1979).
14 Rowe, J.W.; Kilgore, A.; Robertson, G.L.: Evidence in man that cigarette smoking induces vasopressin release via an airway-specific mechanism. J. clin. Endocr. Metab. *51:* 170 (1980).
15 Dunn, F.L.; Brennan, T.J.; Nelson, A.E.; Robertson, G.L.: The role of blood osmolality and volume in regulating vasopressin secretion in the rat. J. clin. Invest. *52:* 3212 (1974).

16 Wade, C.E.; Bie, P.; Keil, L.C.; Ramsay, D.J.: Osmotic control of plasma vasopressin in the dog. Am. J. Physiol. *243:* E287 (1982).

17 Zerbe, R.L.; Robertson, G.L.: Osmoregulation of thirst and vasopressin secretion in human subjects. Effects of various solutes. Am. J. Physiol. *244:* E607 (1983).

18 Rundgren, M.; Eriksson, S.; Appelgren, B.: Urea induced inhibition of antidiuretic hormone secretion. Acta physiol. scand. *106:* 491 (1979).

19 Sklahr, A.H.; Linas, S.L.: The osmolar gap in renal failure. Ann. intern. Med. *98:* 481 (1983).

20 Szatalowicz, V.L.; Goldberg, J.P.; Anderson, R.J.: Plasma antidiuretic hormone in acute respiratory failure. Am. J. Med. *72:* 583 (1982).

21 Szatalowicz, V.L.; Arnold, P.E.; Chaimovitz, C.; Bichet, D.; Schrier, R.W.: Radio-immunoassay of plasma arginine vasopressin in hyponatremic patients with congestive heart failure. New Engl. J. Med. *305:* 263 (1981).

22 Bichet, D.; Szatalowicz, V.L.; Chaimovitz, C.; Schrier, R.W.: Role of vasopressin in abnormal water excretion in cirrhotic patients. Ann. intern. Med. *96:* 413 (1982).

23 Gross, P.; Rascher, W.; Ritz, E.: Die klinische Hyponatriämie. Eine Vasopressin-Störung. Proc. 90. Tag. dt. Ges. Innere Med. 1984, p. 397.

24 Fine, L.G.; Schlöndorff, D.; Trizna, W.; Gilbert, R.M.; Bricker, N.S.: Functional profile of the isolated uremic nephron. J. clin. Invest. *61:* 1519 (1978).

25 Schrier, R.W.; Berl, T.: Disorders of water metabolism; in Schrier, Renal and electrolyte disorders (Little, Brown, Boston 1982).

26 Pastoriza-Munoz, E.; Easterling, R.E.; Malvin, R.L.: The effect of plasma calcium on plasma ADH levels in anephric patients. Nephron *16:* 449 (1976).

27 Nord, E.; Danovitch, G.M.: Vasopressin response in hemodialysis patients. Kidney int. *16:* 234 (1979).

28 Horky, K.; Sramkova, J.; Lachmanova, J.; Tomasek, R.; Dvorakova, J.: Plasma concentration of antidiuretic hormone in patients with chronic renal insufficiency on maintenance dialysis. Hormone metabol. Res. *11:* 241 (1979).

29 Shimamoto, K.; Watarai, I.; Miyahara, M.: Plasma vasopressin in patients undergoing chronic hemodialysis. J. clin. Endocr. Metab. *45:* 714 (1977).

30 Rauh, W.; Hund, E.; Sohl, G.; Rascher, W.; Mehls, O.; Schärer, K.: Vasoactive hormones in children with chronic renal failure. Kidney int. *24:* S27 (1983).

31 Ardaillou, R.; Benmansour, M.; Rondeau, E.; Caillens, H.: Métabolisme et sécrétion de l'hormone antidiurétique dans l'insuffisance rénale, l'insuffisance cardiac, et l'insuffisance hépatique. Actual. néphrologiques Hôpital Necker, Paris 1983.

32 Quillen, E.W.; Skelton, M.M.; Rubin, J.; Cowley, A.W.: Osmotic control of vasopressin with chronically altered volume states in anephric dogs. Am. J. Physiol. *247:* E355 (1984).

33 Monos, E.; Cox, R.H.; Peterson, L.H.: Direct effect of physiological doses of arginine vasopressin on the arterial wall in vivo. Am. J. Physiol. *234:* H167 (1978).

34 Penit, J.K.; Faure, M.; Jard, S.: Vasopressin and angiotensin II receptors in rat aortic smooth muscle cells in culture. Am. J. Physiol. *244:* E72 (1983).

35 Cowley, A.W.; Switzer, S.J.; Guinn, M.M.: Evidence and quantification of the vasopressin arterial pressure control system in the dog. Circulation Res. *46:* 58 (1980).

36 Burnier, M.; Diollaz, J.; Brunner, D.B.; Brunner, H.R.: Blood pressure maintenance in awake dehydrated rats. Renin, vasopressin and sympathetic activity. Am. J. Physiol. *245:* H203 (1983).

37   Wang, B.C.; Sundet, W.D.; Hakumäki, M.O.K.; Goetz, K.L.: Vasopressin and renin responses to hemorrhage in conscious cardiac-denervated dogs. Am. J. Physiol. *246:* H399 (1983).
38   Thames, M.D.; Schmid, P.G.: Cardiopulmonary receptors with vagal afferents tonically inhibit ADH release in the dog. Am. J. Physiol. *237:* H 299 (1979).
39   Möhring, J.; Glänzer, K.; Maciel, J.A.; Düsing, R.; Kramer, H.J.; Arbogast, R.; Koch-Weser, J.: Greatly enhanced pressor response to antidiuretic hormone in patients with impaired cardiovascular reflexes due to idiopathic orthostatic hypotension. J. cardiovasc. Pharmacol. *2:* 367 (1980).
40   Rascher, W.; Schömig, A.; Kreye, V.; Ritz, E.: Diminished vascular response to noradrenaline in experimental chronic uremia. Kidney int. *21:* 20 (1982).
41   Davis, R.; Forsling, M.: Vasopressin release after postural changes and syncope. J. Endocr. *65:* 59 (1975).
42   Wiggins, R.C.; Basar, I.; Slater, J.D.H.; Forsling, M.; Ramage, C.M.: Vasovagal hypotension and vasopressin release. Clin. Endocr. *6:* 387 (1977).
43   Kersh, E.S.; Kronfield, S.J.; Unger, A.; Popper, R.W.; Cantur, S.; Cohn, K.: Autonomic insufficiency in uremia as a cause of hemodialysis induced hypotension. New Engl. J. Med. *290:* 650 (1974).
44   Velez, R.; Woodard, T.; Henrich, W.: Prospective identification of patients with autonomic insufficiency. Kidney int. *24:* 57A (1984).
45   Zerbe, R.L.; Henry, D.P.; Robertson, G.L.: Vasopressin response to orthostatic hypotension. Am. J. Med. *74:* 265 (1983).
46   Ichikawa, I.; Brenner, B.M.: Evidence for glomerular actions of ADH and dibutyryl cyclic AMP in the rat. Am. J. Physiol. *233:* F102 (1977).
47   Allen, J.A.; Roddie, I.C.: The effect of antidiuretic hormone on the rate of sweat production in man. J. Physiol., Lond. *212:* 37P (1971).
48   Taussig, L.M.; Braunstein, G.D.: Effects of vasopressin on sweat rate and composition in patients with diabetes insipidus and normal controls. J. invest. Derm. *60:* 197 (1973).
49   Szczepanska-Sadowska, E.; Sobocinska, J.; Sadowski, D.: Central dipsogenic effect of vasopressin. Am. J. Physiol. *242:* R372 (1982).

PD Dr. Peter Gross, Medizinische Universitätsklinik, Sektion Nephrologie,
Bergheimer Strasse 56a, D–6900 Heidelberg 1 (FRG)

Discussion

Gurland: Thank you, Dr. Gross, for giving us so many important data and also quite a lot of questions.

Koch: What was the renin status of these patients with irreversible hypotension on haemodialysis?

Gross: I agree with you that this is a very important question. We have obtained measurements in these groups of patients during hypotension, but I do not know the results yet.

Ardaillou: To come back to the role of AVP on water retention in patients with hyponatremic chronic renal failure, I think that creation and maintenance of hyponatremia have distinct mechanisms. The mechanism of creation is the diminution of GFR. It is impossible to excrete normally a water load with a low GFR. Thus, the appearance of hyponatremia is related probably to the diminution of GFR. Subsequently, there is an inappropriate response of AVP to hyponatremia. In spite of hyponatremia, high levels of plasma AVP are observed. I believe in agreement with the study of Prof. Schrier (Denver, Colo.) that the role of diuretics in patients with chronic renal failure is not very important. The main role seems to be a severe reduction in GFR. As also shown by Dr. Schrier in patients with nephrotic syndrome, predominately those with low GFR are unable to excrete a water load normally.

Gross: I agree with your first point. There is ADH insensitivity in the uremic nephron. There is also severely restricted glomerular filtration rate in patients with advanced renal failure. This should be sufficient to impair water excretion significantly. As to the second point, a disagreement between our opinions may be that we do not know the role of vasopressin fragments in the apparently high plasma vasopressin concentration of uremia. I think this problem needs to be answered. We intend to study it by HPLC separation. At the present time, there are no hard data to base any judgement on, we can only suspect.

Dobbelstein: Did your hypotensive episodes react favourably to treatment with vasopressin?

Gross: This is a tempting option which I have had in my mind, too. I think as long as this hypotension is not properly clarified, I do not wish to speculate on this point. The literature records potential side-effects of vasopressin treatment. When vasopressin was given to suppress gastrointestinal bleeding, there are more than 30 reports in the literature over the last 8 or 9 years reporting gangrene of an extremity, myocardial infarction and stroke. Now when you consider that most uremic patients are likely to suffer some degree of cardiovascular arteriosclerosis then these reports assume additional importance. On the other hand, you can say that — and this was indeed observed — some patients will have excessive endogenous vasopressin concentration during nausea anyway. Some levels were as high as 40 pg/ml and nothing seems to have happened to these patients in terms of side-effects. I think this issue remains to be considered in the future.

Wizemann: On your first slide you mentioned the problem of thirst. Could you expand a little bit on thirst and its connection to ADH?

Gross: This is an interesting issue. There was a single paper from Poland in the *American Journal of Physiology* 3 years ago reporting that the injection of vasopressin causes thirst in the dog. As we know many of our patients have hyperosmolality. Many of our uremic patients also have excessive thirst and measurable vasopressin concentrations are elevated. One can therefore speculate whether this contributes to their thirst. This observation has been reported only in one paper. I am in no position to judge this issue. Perhaps the clinical availability of vasopressin antagonists in the next few years, may help us to answer whether vasopressin contributes to our patients' thirst.

Ritz: Could you please comment on the interrelation between age and hyponatremia? You are well aware of studies which found a surprisingly high incidence of hyponatremia, but almost all in elderly patients acutely on diuretics. The cases that we saw of hyponatremia in advanced renal failure were also over 60 years of age. Could you briefly comment on the mechanism and the cause?

Gross: I know of two reports in the literature that show an increase of vasopressin concentrations with age, but most of these patients were between 70 and 80 years old, some of these patients manifested otherwise unexplained hyponatremia. It is thought that receptors in the aortic arch, the carotid bodies, and perhaps the left and right atrium, cause tonic inhibition of hypothalamic vasopressin secretion. With advancing age and structural changes in arterial walls and autonomic nerves this suppressive influence may diminish. Thereby vasopressin secretion may be less inhibited than before.

Ritz: This seems like a logical explanation. Do you have any information, however, on Alzheimer's disease and related syndromes of disturbed cholinergic innervation? Is hyponatremia more common?

Gross: I do not know that this problem has been studied. I only know that there are studies of vasopressin treatment in Alzheimer's disease. In general, inappropriate vasopressin stimulation seems to be more common in a general psychiatric population.

Bommer: Dr. Gross, you have discussed very cautiously the influence of vasopressin and blood pressure in dialysis patients. A year ago I had discussions with Dr. Brodde and in consequence we tried, in 3 patients who were hypotensive, 1 h after the beginning of dialysis, to give them DDAVP and the blood pressure did not change.

Gross: I am sorry to say that DDAVP is not a good candidate for this purpose.

Bommer: Yes, that is true.

Gross: DDAVP is a specific synthetic analogue with predominant hydro-osmotic activity and very limited vasoactive activity. The latter, however, is required in this context. One should therefore use arginine-vasopressin or lysine-vasopressin or one of many other analogues with pressor activity. So with DDAVP it is not surprising you did not observe an increased blood pressure.

Heidland: We have treated one hemodialysis patient suffering from severe autonomic neuropathy who resisted all manoeuvres, with vasopressin infusion during each hemodialysis over 1 year. She tolerated hemodialysis treatment very well and even after the end of infusion, probably due to elevated vasopressin concentration, the blood pressure was well maintained.

Ritz: This is not specific, of course. It has been described in the mid-1950s that this is the superior treatment.

Gurland: Any further questions?

Ritz: As an added consideration to what Gross said. This is, of course, that we must be extremely aware of the dangers. One danger I would point out is mesenteric vasoconstriction which is really what I am concerned about apart from coronary. There have been reports of cases on hemodialysis who had mesenteric infarction after digitalis and this is one complication anticipated if one uses drugs indiscriminately.

Contr. Nephrol., vol. 50, pp. 64–72 (Karger, Basel 1986)

Thyroid Function in Renal Failure

Elaine M. Kaptein

Division of Nephrology, University of Southern California, Los Angeles, Calif., USA

Patients with acute or chronic renal failure may have alterations in their serum levels of thyroid hormones secondary to changes in thyroid hormone metabolism induced by the nonthyroidal illness and/or abnormal thyroid gland function. Further, the clinical as well as the biochemical features of hyperthyroidism or hypothyroidism may be mimicked or masked by the concurrent renal disease. Hyperthyroidism and hypothyroidism induce significant morbidity and potential mortality, and, if recognized, are totally reversed by appropriate therapy. Thus, differentiating the changes in serum thyroid hormone levels due to the nonthyroidal illness from those secondary to thyroid dysfunction, and recognizing the effects of illness on the biochemical features of hyper- and hypothyroidism are an integral part of the evaluation and management of patients with renal failure.

Thyroid Hormone Metabolism in Euthyroid Patients with Renal Failure

Patients with acute [1] or chronic [2] renal failure frequently have reduced serum levels of total T_4 and T_3 and decreased free T_4 and T_3 index values, as in other nonthyroidal illnesses [3]. The free T_4 index values are misleadingly low since free T_4 levels by equilibrium dialysis are usually normal in these patients [2] and T_4 production rates are normal in the presence of low total T_4 and free T_4 index values in patients with nonthyroidal illness [4]. Serum concentrations of TBG are normal or minimally decreased [2, 3] in these patients indicating the presence of a reduced binding affinity of TBG for T_4. Basal serum TSH levels are normal or minimally increased and the

TSH response to TRH is normal or blunted, as in other nonthyroidal illnesses [1-3].

Serum levels of reverse T_3, the calorogenically inactive thyroid hormone, are usually normal in patients with acute [3] as well as those with chronic [2, 3] renal failure. This is in contrast to the elevated reverse T_3 concentrations observed in the majority of patients with nonrenal illnesses [3]. This difference may be related to the state of secondary hyperparathyroidism present in patients with renal failure [5, 6]. The observation that serum reverse T_3 levels also fail to increase in patients with nephrotic syndrome who have normal glomerular filtration rates [7] and secondary hyperparathyroidism [8], and in those with primary hyperparathyroidism and normal renal function [9], in the presence of reduced serum levels of T_3, supports this possibility. The normal serum levels of reverse T_3 in chronic renal failure appear to be due to a shift of the hormone from the vascular to the extravascular tissues [10] secondary to increased reverse T_3 transport out of serum [11].

Kinetic studies in patients with chronic renal failure indicate that thyroid gland production of T_4 and T_3 is normal, while the peripheral conversion of T_4 to T_3 is reduced [12]. In addition, the rate of T_4 transport from serum to extravascular tissues is significantly reduced [10]. This may be primarily due to impaired T_4 transport into rapidly equilibrating tissues such as liver [13]. This reduction of T_4 entry into tissues may contribute to the decrease in T_3 production from T_4.

Although T_3 is believed to be the most metabolically active thyroid hormone, these patients are clinically euthyroid as judged by the normal clinical index score, basal metabolic rate, Achilles deep tendon reflex relaxation time, and systolic time interval [14]. The reduced serum levels of free T_3 and decreased T_3 production rates may represent an adaptive phenomenon to minimize the catabolic effects of renal failure. Studies in the uremic rat suggest the presence of an heterogeneous tissue response to renal failure; the nuclear content of T_3 is reduced in the liver but normal in the pituitary gland in the presence of low circulating T_3 levels in these animals [15].

The alterations in the serum levels, distribution and metabolism of T_4 and T_3 in patients with renal failure are similar to those in nonrenal illnesses. A number of common factors may underly these changes which include the general state of nonthyroidal illness and the associated malnutrition and negative nitrogen balance. The direct correlations between serum levels of T_3 and those of albumin and transferrin in patients with chronic renal failure [16] as well as in nonrenal illnesses [16] support this possibility. Further, the magnitude of the changes in serum T_4 and T_3 levels may relate to the severity of the nonthyroidal illness. This is evidenced by the finding that serum T_4 and T_3 levels are lower in patients with acute renal failure complicated by nonrenal illnesses than in those with acute renal failure alone [1]. As expected, serum thyroid hormone levels return to normal as the renal function recovers in these patients [1].

Thyroid Dysfunction in Patients with Renal Failure

The prevalence of goitre in patients with end-stage chronic renal failure varies from 58% in Utah to 0% in London [2]. This variability may be related to geographic and genetic factors. In 255 patients with end-stage chronic renal failure studied in our institution, 35% had significant thyroid gland enlargement compared to 11% in hospitalized patients without evidence of renal disease [16]. In addition, our patients with end-stage chronic renal failure had a higher prevalence of hypothyroidism but a similar frequency of hyperthyroidism as the general population [16]. Thus, some factor(s) associated with chronic uremia may induce goitre and hypothyroidism in uremic patients predisposed to developing thyroid dysfunction.

Differentiation of Thyroidal from Nonthyroidal Disorders in Patients with Renal Failure

Many clinical manifestations of hypothyroidism may resemble those of chronic uremia. These include pallor, puffiness, fatigue, lethargy, constipation, and cold intolerance. Further, pericardial and pleural effusions may occur in both disorders. However, the clinical significance of the pericardial effusion differs since tamponade occurs frequently in uremia and rarely in hypothyroidism. Similarly, signs and symptoms of hyperthyroidism may be masked by or attributed to the underlying renal failure, other concurrent illnesses and/or the pharmacologic agents administered. Many patients with renal failure have underlying cardiac disease with atrial fibrillation, angina pectoris and/or congestive heart failure, which could be exacerbated by hyperthyroidism. Thus, the diagnosis of thyroid dysfunction in patients with renal failure should be primarily based on biochemical detection and confirmation.

The serum total T_4 and free T_4 index values are the most widely used screening tests for hyper- or hypothyroidism. Elevated or reduced values

Table I. Evaluation scheme for patients with renal failure and decreased free T_4 index values

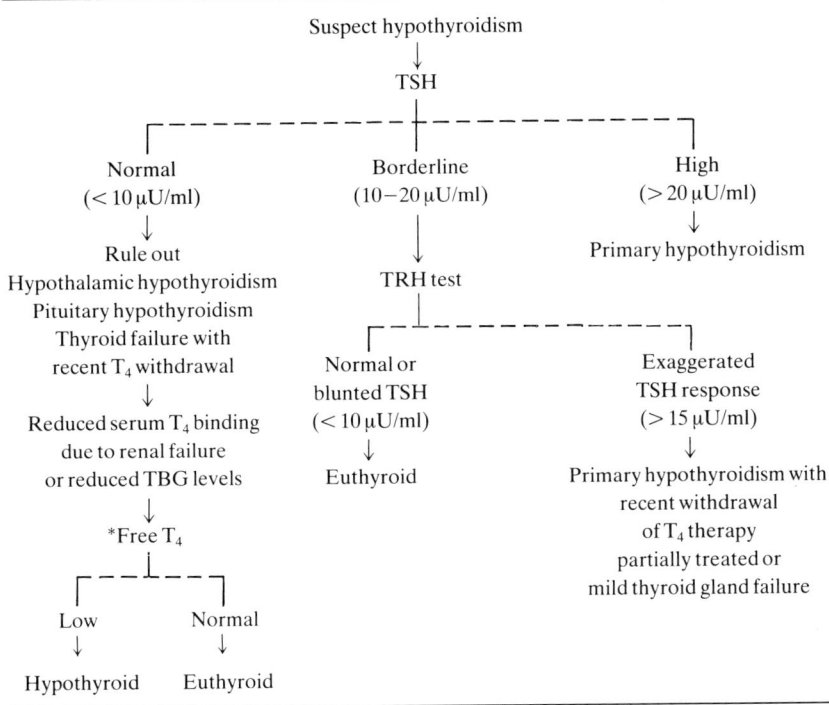

* Free T_4 by equilibrium dialysis or Clinical Assays 2-step method.

may result from nonthyroidal illnesses [3, 17] as well as thyroid gland dysfunction, therefore, additional tests are required to establish a diagnosis. A schematic approach to this evaluation in patients with renal failure is depicted in tables I and II. If the free T_4 index value is reduced, hypothyroidism must be ruled out; a serum TSH level is the next most appropriate test. If the serum TSH level is increased to above 20 μU/ml, the diagnosis is hypothyroidism due to thyroid gland failure and therapy with *L*-thyroxine should be initiated. Since euthyroid patients with renal failure frequently have reduced free T_4 index values due to impaired T_4 binding to serum carrier proteins, the appropriate dose of *L*-thyroxine should be determined based on clinical symptoms and signs and by monitoring serum TSH levels. The maintenance dose of blood *L*-thyroxine should not exceed 0.15 mg/day in young patients or 0.10 mg/day in those over the age of 50 years unless the

Table II. Evaluation scheme for patients with renal failure and increased free T$_4$ index values

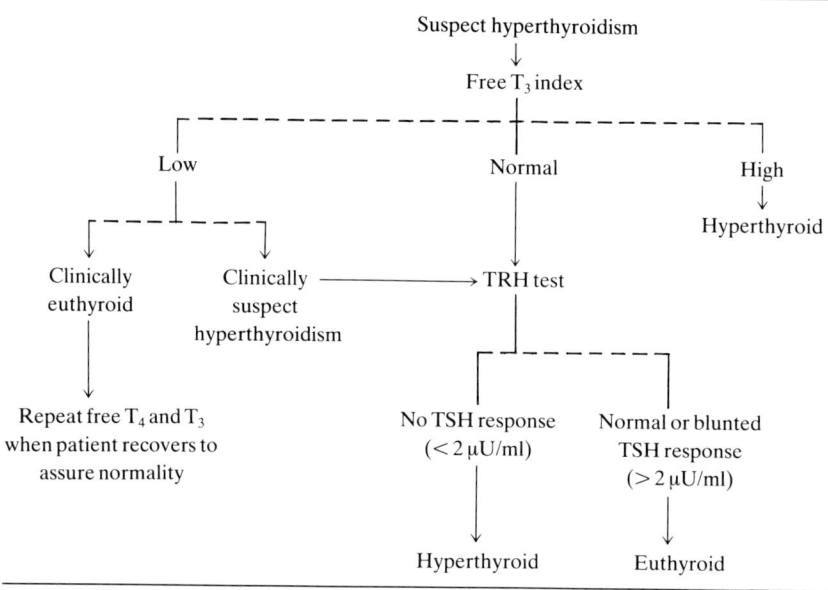

TSH value remains above $10-15\,\mu$U/ml. Since the half-life of T$_4$ is usually $7-10$ days, serum TSH levels should not be checked sooner than $4-6$ weeks after the *L*-thyroxine dose is initiated or changed to assure stable free T$_4$ values have been attained.

Serum TSH values between 10 and 20 μU/ml in the presence of low free T$_4$ index values may be observed in euthyroid patients with renal failure as well as in hypothyroid patients with mild or partially treated thyroid gland failure or in those who have been recently withdrawn from *L*-thyroxine therapy. In these patients a TRH test may be of value; a TSH response to TRH of less than 10 μU/ml is compatible with euthyroidism while an increase in TSH of more than 15 μU/ml is indicative of primary hypothyroidism and requires *L*-thyroxine therapy.

A normal or minimally elevated serum TSH concentration ($<10\,\mu$U/ml) in the presence of a reduced free T$_4$ index value requires exclusion of hypothyroidism due to hypothalamic or pituitary lesions. These disorders are frequently associated with multiple endocrine deficiencies which may be evident by history and physical examination; the presence of hypotension warrants evaluation of adrenal cortical status by measuring a serum cortisol level. A normal serum TSH level with a low free T_4 index value may also be observed for 4–6 weeks following withdrawal of L-thyroxine therapy in a patient with thyroid gland failure. If these causes of hypothyroidism are excluded, reduced serum T_4 binding due to the non-thyroidal illness or to decreased serum TBG concentrations is most likely; these patients are euthyroid and should have a normal free T_4 level by equilibrium dialysis or the Clinical Assays 2-step method [18]. A sharp de-marcation of free T_4 levels determined by these two methods may not be present between patients with mild to moderate hypothyroidism and euthy-roid patients with the low total T_4 state of nonthyroidal illness [18]. Further, most other free T_4 methods give misleadingly low free T_4 estimates in euthyroid patients with the low T_4 state of illness [18]. Thus, serum free T_4 values must be interpreted with caution in patients with renal failure and used as a confirmatory rather than as a definitively diagnostic test.

An elevated free T_4 index value may be secondary to hyperthyroidism or to a mild to moderate nonthyroidal illness such as renal failure. Although high free T_4 index values are much less commonly observed than reduced values in chronic renal failure, they occur with a frequency similar to that of hyperthyroidism [16]. The serum total and free T_4 levels are increased in hyperthyroidism secondary to enhanced T_4 production by the thyroid gland; the TSH response to TRH is absent. In contrast, in euthyroid pa-tients with the high T_4 state of illness, the elevated total and free T_4 levels result from reduced clearance of T_4 from serum while production rates of the hormone are normal [19]; the TSH response to TRH is normal or blunted rather than absent [17].

As indicated in table II, a serum free T_3 index value may be of assis-tance in differentiating euthyroid from hyperthyroid patients with high free T_4 index values [17]. An elevated free T_3 index value indicates hyperthyroid-ism, however, a normal or reduced value does not exclude this diagnosis since extrathyroidal T_3 production is decreased by illness. If the free T_3 index value is within the normal range, a TRH test is indicated. A TSH response to TRH which is absent or less than 2 µU/ml indicates hyper-thyroidism while a normal or blunted response excludes this diagnosis. In severely ill patients with a reduced free T_3 index value and intractable con-gestive heart failure, angina, tachycardia, or other clinical evidence to suggest hyperthyroidism, a TRH test should be performed. In the absence of clinical suspicion of hyperthyroidism, the free T_4 and T_3 index values should be repeated when the patient's illness improves to assure normality. In hyperthyroid patients with concurrent critical illness, the free T_4 index may be reduced to normal or low levels; the diagnosis of hypothyroidism must be suspected clinically and confirmed by the absence of a TSH response to TRH. The prompt initiation of specific therapy for the hyperthyroid patient with renal failure is essential to prevent thyroid storm and to alleviate otherwise unresponsive cardiac failure, coronary insufficiency and/or tachyarrhythmias.

References

1   Kaptein, E.M.; Levitan, D.; Feinstein, E.I.; Nicoloff, J.T.; Massry, S.G.: Alterations of thyroid hormone indices in acute renal failure and in acute critical illness and without acute renal failure. Am. J. Nephrol. *1:* 138−143 (1981).

2   Kaptein, E.M.; Feinstein, E.I.; Massry, S.G.: Thyroid hormone metabolism in renal diseases. Contr. Nephrol., vol. 33, pp. 122−135 (Karger, Basel 1982).

3   Wartofsky, L.; Burman, K.D.: Alterations in thyroid function in patients with systemic illness. The 'euthyroid sick syndrome'. Endocr. Rev. *3:* 164−217 (1982).

4   Kaptein, E.M.; Robinson, W.J.; Grieb, D.A.; Nicoloff, J.T.: Peripheral serum thyroxine, triiodothyronine and reverse triiodothyronine kinetics in the low thyroxine state of acute nonthyroidal illnesses. A noncompartmental analysis. J. clin. Invest. *69:* 526−535 (1982).

5   Massry, S.G.; Coburn, J.W.; Peacock, M.; Kleeman, C.R.: Turnover of endogenous parathyroid hormone in uremic patients and those undergoing hemodialysis. Trans. Am. Soc. artif. internal Organs *18:* 416−420 (1972).

6   Massry, S.G.; Arieff, A.I.; Coburn, J.W.; Palmieri, G.M.; Kleeman, C.R.: Divalent ion metabolism in patients with acute renal failure; studies on mechanisms of hypocalcemia. Kidney int. *5:* 437−445 (1974).

7   Feinstein, E.I.; Kaptein, E.M.; Nicoloff, J.T.; Massry, S.G.: Thyroid function in patients with nephrotic syndrome and normal renal function. Am. J. Nephrol. *2:* 70−76 (1982).

8   Goldstein, D.A.; Haldimann, B.; Sherman, D.; Norman, A.W.; Massry, S.G.: Vitamin D metabolites and calcium metabolism in patients with nephrotic syndrome and normal renal function. J. clin. Endocr. Metab. *52:* 116−121 (1981).

9   Kaptein, E.M.; Massry, S.G.; Quion-Verde, H.; Singer, F.R.; Feinstein, E.I.; Nicoloff, J.T.; Sharp, C.F.: Serum thyroid hormone indicies in primary hyperparathyroidism. Archs intern. Med. *144:* 313−315 (1984).

10 Kaptein, E.M.; Feinstein, E.I.; Nicoloff, J.T.; Massry, S.G.: Serum reverse triiodothyronine and thyroxine kinetics in patients with chronic renal failure. J. clin. Endocr. Metab. *57:* 181−189 (1983).

11 Kaptein, E.M.; Chang, E.; Feinstein, E.I.; Nicoloff, J.T.; Massry, S.G.: Extravascular binding and transport of reverse T_3 in critical nonrenal nonthyroidal illnesses (CI) and in chronic renal failure (CRF). Proc. 59th Ann. Meet. Am. Thyroid Ass., New Orleans 1983, p. T-35.

12 Lim, V.S.; Fang, V.S.; Katz, A.I.; Refetoff, S.: Thyroid dysfunction in chronic renal failure: a study of the pituitary-thyroid axis and peripheral turnover kinetics of thyroxine and triiodothyronine. J. clin. Invest. *60:* 522−534 (1977).

13	Kaptein, E.M.; Chang, E.; Feinstein, E.I.; Massry, S.G.; Nicoloff, J.T.: Localization of reduced extravascular thyroxine binding in nonthyroidal illness. Proc. 58th Ann. Meet. Am. Thyroid Ass., Quebec 1982, p. T-5.
14	Spector, D.A.; Davis, P.J.; Helderman, J.H.; Bell, B.; Utiger, R.D.: Thyroid function and metabolic state in chronic renal failure. Ann. intern. Med. *85:* 724–730 (1976).
15	Lim, V.S.; Passo, C.; Murata, Y.; Ferrari, E.; Nakamura, H.; Refetoff, S.: Reduced triiodothyronine content in liver but not pituitary of the uremic rat model. Demonstration of changes compatible with thyroid hormone deficiency in liver only. Endocrinology *114:* 280–286 (1984).
16	Quion-Verde, H.; Kaptein, E.M.; Chooljian, C.J.; Rodriguez, H.J.; Massry, S.G.: Prevalence of thyroid disease in chronic renal failure (CRF) and dialysis patients. Proc. 9th Int. Congress Nephrol., Los Angeles 1984, p. 120A.
17	Gavin, L.A.; Rosenthal, M.; Cavalieri, R.R.: The diagnostic dilemma of isolated hyperthyroxinemia in acute illness. J. Am. med. Ass. *242:* 251–253 (1979).
18	Kaptein, E.M.; MacIntyre, S.S.; Weiner, J.M.; Spencer, C.A.; Nicoloff, J.T.: Free thyroxine estimates in nonthyroidal illnesses: comparison of eight methods. J. clin. Endocr. Metab. *52:* 1073–1077 (1981).
19	Bianchi, R.; Mariani, G.; Molea, N.; Vitek, F.; Cazzuola, F.; Carpi, A.; Mazzuca, C.; Toni, M.G.: Peripheral metabolism of thyroid hormones in man. I. Direct measurement of the conversion rate of thyroxine to 3,5,3′-triiodothyronine (T_3) and determination of the peripheral and thyroidal production of T_3. J. clin. Endocr. Metab. *56:* 1152–1163 (1983).

Elaine M. Kaptein, MD, University of Southern California, Department of Medicine, Division of Nephrology, School of Medicine, 2025 Zonal Avenue, Los Angeles, CA 90033 (USA)

Discussion

Kokot: First of all, I would like to congratulate you for your excellent presentation. As you know, in critically ill patients usually there exsists a negative relationship between the T_3 and rT_3 plasma levels. We, too, in our acute renal failure patients found this negative relationship. From your figures I noticed that normal rT_3 levels were found in patients with acute renal failure. Could you speculate about the mechanism of this finding?

Kaptein: We have done reverse T_3 kinetic studies in a small group of patients with acute renal failure during the oliguric phase and with recovery. These patients were very much like chronic renal failure patients. They have normal reverse T_3 production rates and a shift of reverse T_3 from the vascular to the extravascular compartment with an acceleration of the fractional transport rate. This is also seen in chronic renal failure. We did not observe any inverse relationship between T_3 and rT_3 levels in these patients and I do not understand why our populations differ.

Kokot: I have a further question on the same point. What happened to the reverse T_3 which was shifted to the extravascular compartment?

Kaptein: This is a postulation which we are in the process of proving or refuting by kinetic analysis of external quantitation of tissue uptake of rT_3. The postulate that I had initially from serum data alone was that the rate of reverse T_3 transport into tissues was accelerated into both rapidly and slowly equilibrating tissues and that the extravascular tissue binding was also enhanced. However, when you use serum data and you do compartmental analysis it is purely speculative. Subsequent to that, we have done external quantitation, and preliminary data suggest that the quantity of rT_3 in liver is definitely increased, and the rate at which it accumulates is increased. However, we looked at muscle as an example of the slowly equilibrating tissues and we do not see these changes in that compartment.

Ritz: Coming back to the TSH levels, which are more frequently elevated than in the general population, could you comment on whether your patients were on vitamin D treatment or not. The reason I am asking is that in the hypophysis receptors are present for $1,25(OH)_2D_3$ and you are certainly aware of recent reports that $1,25(OH)_2D_3$ increases the amount of TSH released by TRF. Could you comment on this?

Kaptein: Very few of our chronic renal failure patients, perhaps 2 or 3 of the total group, were on vitamin D supplements. The elevations in TSH were mild to moderate and were usually between 5 and 10 μU/ml. This elevation is very common also in nonrenal patients. Recently, in the *Journal of Clinical Endocrinology and Metabolism* [Warner, B.A., et al., *60:* 263, 1985] patients with systemic illnesses were reported to have an increase of immunoreactive LH that was not bioactive; this is also strongly suspected for TSH, since any state that increases bioactive basal TSH levels also induces an increase in the TSH response to TRH and usually implies a decrease in pituitary TSH suppression. These patients with nonthyroidal illnesses are paradox if this elevation is bioactive TSH because the baseline levels are elevated but the response to TRH is blunted. Therefore, I suspect that they may have an increase of immunoreactive TSH that is not bioactive.

Bommer: I have a question concerning T_4 binding. The TBG is only slightly lowered, albumin is normal in most patients and the prealbumin is not markedly changed. Is there a disturbance of the affinity between T_4 and TBG? Have you any information in this field?

Kaptein: TBG concentrations are minimally altered. TBPA levels are decreased by approximately 50% but they bind a very small faction of T_4, and albumin levels are not usually decreased that much, and they bind a very small fraction. To the point of the affinity of TBG for T_4, T_3 and rT_3, this has been studied extensively by Chopra and his co-workers at UCLA. Chopra has been the primary person working on the presence of a circulating inhibitor which he believes is derived from tissues, probably in the process of tissue breakdown. They have shown that there is a decreased affinity of TBG for T_4 in patients with a variety of nonthyroidal illnesses. He has now homogenized normal rat tissue and if he puts that into the serum it has the same characteristics as the serum inhibitor and may be a lipoprotein. An increase of oleic acid or other free fatty acids may be altering TBG affinity for T_4, T_3 and rT_3.

Koch: Thank you for your very thorough review.

Contr. Nephrol., vol. 50, pp. 73–95 (Karger, Basel 1986)

Parathyroid Hormone Secretion
Molecular Events and Regulation

Lynn Helena Caporale, Michael Rosenblatt[1]

Merck Sharp & Dohme Research Laboratories, West Point, Pa., USA

In normal physiology, parathyroid hormone (PTH) serves an important homeostatic role (through its effects on kidney, bone, and gut) in the regulation of extracellular calcium levels within the narrow limits required for optimal functioning of nerve, muscle, and enzymes, and of other hormones [1].

If the intracellular mechanisms and extracellular signals which regulate PTH secretion are not functioning properly, there are serious medical consequences. As we better understand these processes, we will be better able to treat disorders of calcium homeostasis without resorting to surgery.

Within the past several years, progress has been made toward understanding the regulation of PTH secretion, a topic that has been extensively reviewed [2, 3]. In addition to calcium, the principal regulator of PTH secretion, several other secretagogues or modulators of PTH secretion have been identified. The pathophysiology of PTH secretion in hyperparathyroidism has also received extensive recent investigation. A brief review of the normal and abnormal control of PTH secretion follows.

Potential Control Points in the Regulation of PTH Levels: Biosynthesis, Intracellular Processing and Transport, and Packaging of PTH

Like other secreted proteins, the PTH gene encodes a larger precursor form, pre-proparathyroid hormone (Pre-ProPTH) [4–7]. The genomic DNA encoding Pre-ProPTH is transcribed. RNA is processed into mature

[1] We would like to thank Sandra Camburn for expert secretarial assistance.

mRNA, which is translated into the nascent hormone; this in turn undergoes a series of post-translational modifications.

Pre-ProPTH is 31 amino acids longer than PTH. These 31 amino acids are located at the amino terminus. The first 25 amino acids comprise the 'signal sequence', which is responsible for providing the signal that PTH is to be a secreted, rather than cytoplasmic, protein. As this sequence is synthesized on the ribosome, it binds to a 'signal recognition particle' (SRP), stopping translation until the complex (ribosome, SRP, message and signal sequence) binds to a 'docking protein' found on the membrane of the endoplasmic reticulum. As translation then continues, nascent Pre-ProPTH is translocated across the membrane of the rough endoplasmic reticulum (RER), destined for secretion and/or storage. However, it is important to note that proteolytic cleavage and removal of the signal sequence occurs shortly after synthesis resumes on the endoplasmic reticulum and prior to completion of biosynthesis of the entire nascent protein.

In an in vitro model system, a synthetic peptide, corresponding to the precursor-specific (Pre-Pro) sequence of PTH, inhibits processing (translocation and cleavage) not just of Pre-ProPTH, but also of other secreted proteins and peptides (e.g. pre-prolactin and pre-growth hormone) indicating that these other peptides, although synthesized by other tissues, interact in those tissues with a cellular secretory apparatus and recognition sites that are very similar or identical [8, 9]. One might therefore expect that these other proteins have identical, or highly homologous, signal sequences. However, this is not the case. In fact, there is no primary sequence homology among signal sequences. Rather, these sequences are thought to be recognized because of secondary structural features that they share: a highly hydrophobic region, usually containing considerable secondary structure (such as a helical conformation), followed by polar amino acids forming a beta-bend, and terminating at the carboxyl end with a small amino acid, such as glycine or alanine [7, 10]. Cleavage and removal of the pre sequence occurs just beyond this bend, at the carboxyl side of the small amino acid.

The function of the 6-amino acid, highly basic pro sequence of ProPTH, is unknown. Following biosynthesis, the prohormone is transported through the cisternal channels of the RER to the Golgi apparatus, where ProPTH is converted into PTH. This conversion can be inhibited by the microtubule-disrupting agent, colchicine [11]. In addition, it has recently been reported [12] that a phosphorylated form of PTH can be extracted from cultured parathyroid tissue (although there is as yet no evidence that

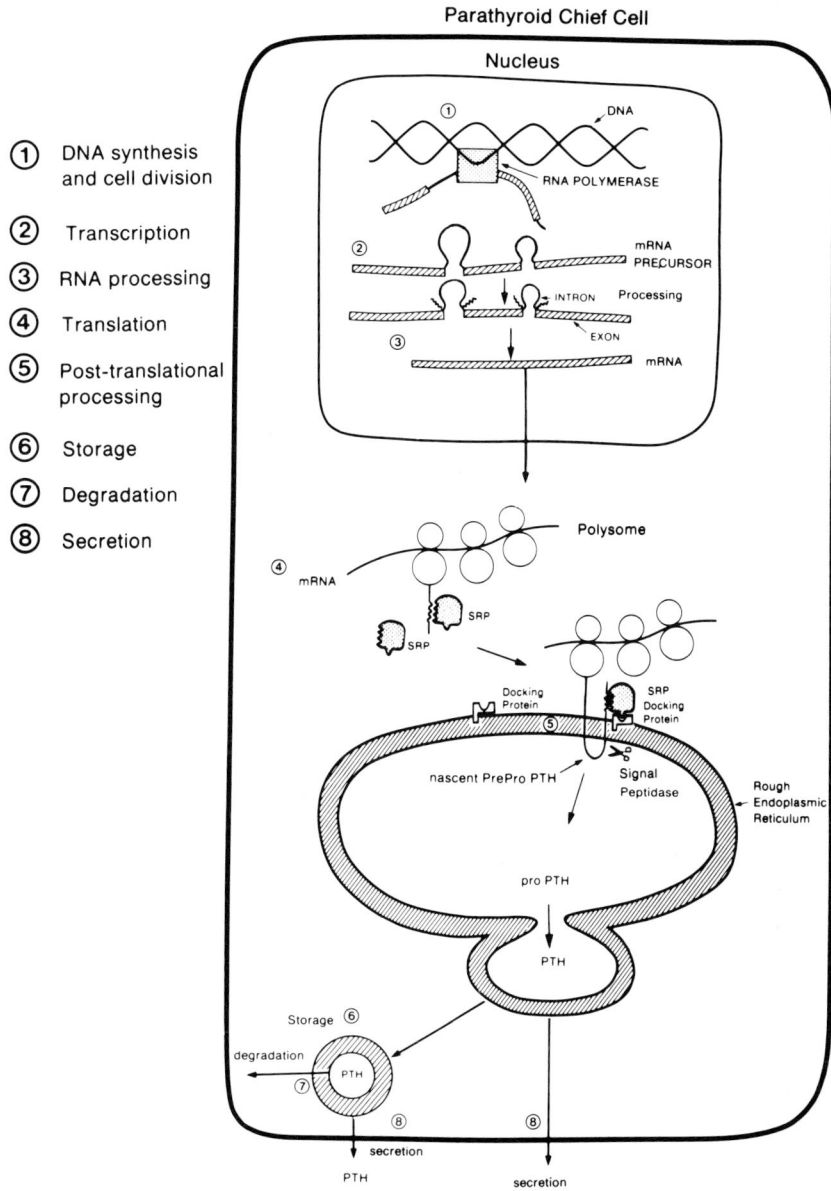

Fig. 1. Schematic representation of major points at which the level of PTH biosynthesis and secretion could be regulated.

this form of PTH occurs in the circulation). PTH is then either released into the circulation, or packaged in secretory vesicles that migrate toward the cell periphery en route to release of the hormone [13–15].

cDNA encoding human Pre-ProPTH was recently introduced into the genome of a rat pituitary cell line. These cells synthesize and process PTH normally, and secrete PTH in response to thyrotropin-releasing hormone (TRH), a normal secretagogue for pituitary cells, not parathyroid cells. The fact that human Pre-ProPTH was properly processed in a rat tissue other than the parathyroid gland, and secreted upon stimulation, illustrates that high homology must exist between the protein sorting and secretory processes in different tissues [16].

Each of the steps illustrated in figure 1 represents a potential control point in the regulation of PTH biosynthesis and ultimate secretion by cells of the parathyroid gland. The major steps to consider are: modulation of the levels of mRNA, the efficiency of transcription and translation, the conversion of ProPTH to PTH, and the proportion of PTH which is immediately secreted instead of stored in secretory granules (where some portion of the stored hormone appears to undergo intracellular degradation). Separately or jointly, these steps determine the amount of hormone available for release by a secretagogue. In addition, different mechanisms may exist for acute (minutes), sustained (hours, days), and chronic (weeks, months) regulation of PTH levels, although most research in this area has examined only short-term effects.

Another major product of the parathyroid glands, called 'parathyroid secretory protein' (PSP), $M_r \simeq 70,000$ is phosphorylated and released in a calcium-regulated manner [17, 18]. Although the function of PSP, which is found together with PTH in secretory granules, is unknown, it is thought that it might have a structural, functional, or regulatory role in the secretory process, since a highly homologous protein has been found in secretory granules of other tissues [19].

Regulation of PTH Secretion

The chief cell is the principal cell responsible for biosynthesis of PTH. After biosynthesis, PTH may be released immediately, or may remain intracellular until it is released upon a physiologic stimulus or other secretagogue. Although ProPTH accounts for < 2% of the hormone secreted [20], parathyroid neoplasms may secrete higher proportions of PTH precursor forms [21]. In addition, the intact hormone may undergo partial degradation intracellularly, leading to the secretion of hormone fragments. Although the Golgi apparatus produces large numbers of immature secretory vesicles, the parathyroid cell contains relatively few mature membrane-enclosed secretory granules, with only small quantities of stored hormone in comparison with other protein-secreting cells (succient hormone for only 7 h of normal or 1.5 h of maximal secretion) [1]. More than one pool of PTH appears to exist intracellularly, and regulation of secretion of these pools differs. Newly synthesized hormone appears to be secreted by a 'bypass pathway', whereas stored hormone is released via the conventional pathways involving mature secretory granules [22, 23].

Calcium

Calcium is the principal physiologic regulator of PTH secretion. While calcium facilitates the tropic secretion of many other hormones and proteins, it acts in an opposite manner on PTH secretion. High calcium levels suppress PTH secretion and low levels stimulate secretion, implying the presence of a unique secretion mechanism or apparatus for PTH (which may be shared by calcium inhibition of renal release of renin) [24]. While the dependence of radioimmunoassay data upon antisera which do not recognize the biologically active amino-terminal region of PTH complicates analysis of the regulation of PTH secretion, data accumulated in multiple in vitro and in vivo systems using different PTH radioimmunoassays have permitted a general formulation for the control of PTH secretion by calcium to emerge. The studies of Mayer [25] (fig. 2), performed by direct cannulation of the venous drainage of parathyroid glands in calves, demonstrated a linear response of PTH secretion by parathyroid cells over only a narrow range of calcium concentration. At calcium levels only slightly below the normal range (below 8 mg/dl), PTH secretion increases dramatically. As calcium levels decline further, PTH secretion rapidly reaches a maximal rate, with no further increase. At high calcium levels (12–18 mg/dl), suppression of PTH secretion is incomplete, even when such levels are maintained for 24 h.

When incubated in a low calcium environment, parathyroid cells secrete increased quantities of PTH, as required for calcium homeostasis. The parathyroid gland is thought to respond to long- and short-term fluctuations in extracellular calcium concentration by different mechanisms.

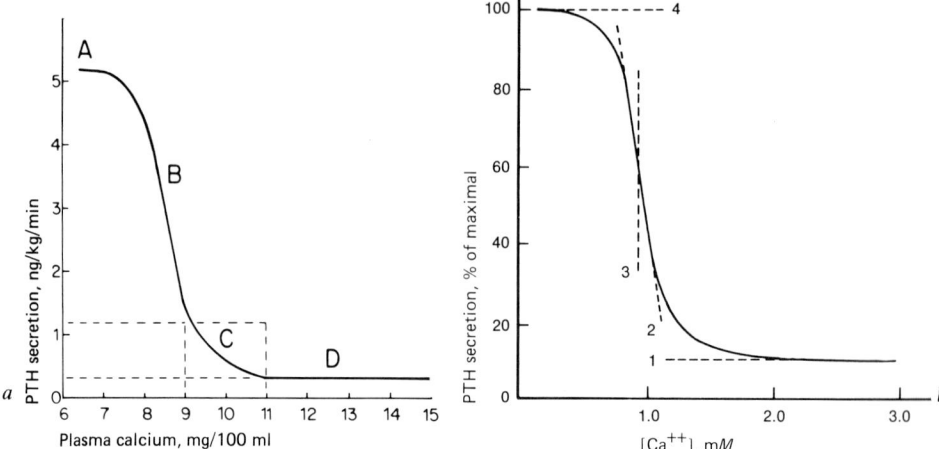

Fig. 2. a Schematic representation of relation of PTH secretion rate to plasma calcium based on data obtained from studies on calves. Parathyroid venous effluent was collected from anesthetized calves during periods of induced changes in blood calcium. PTH was measured by radioimmunoassay. Four separate regions are represented on the curve by the letters A–D. Region C lies within the range of physiologic regulation; region B reflects the rapid increase in secretion rate that occurs in response to hypocalcemia, reaching a maximum at a plasma calcium of approximately 7 mg/100 ml. A low but substantial secretion of PTH persists, despite increasing hypercalcemia (region D). Dashed lines indicate range of normal basal plasma calcium and PTH concentrations. *b* Factors which interact to determine PTH levels. The amount of PTH secreted can be thought of as a result of (1) the basal level of PTH secretion; (2) the 'set point'; (3) the steepness of the curve with which calcium levels affect PTH secretion, and (4) the maximal rate of secretion possible.

Significantly more mRNA encoding for PTH is found in a population of parathyroid cells incubated for several days in 0.5 mM, than in 3 mM, calcium [26]. It is not yet known whether low Ca^{++} increases message levels by increasing the rate of PTH gene transcription in each cell, by recruiting new cells to become PTH secretors, or by decreasing the rate of mRNA degradation. Upon exposure to calcium for shorter periods, the quantity of mRNA does not change detectably.

Although the amount of PTH secreted from parathyroid tissue is much higher in an environment of low, than of high, calcium, in vitro rates of biosynthesis of PTH are only slightly influenced by calcium over periods of hours [15]. Translational efficiency is only minimally influenced by calcium levels, and the long half-life [27] of Pre-ProPTH mRNA makes transcriptional control an unlikely point for acute regulation of PTH biosynthesis. Furthermore, the rates of appearance and disappearance of ProPTH molecules are not altered over the course of hours under these different conditions. Hence, over short time periods, the regulation of secretion vs. storage of PTH by calcium appears to determine levels of PTH in circulation, and this control step occurs after the formation of ProPTH. Chronic stimulation of PTH secretion by hypocalcemia may ultimately increase mRNA levels; another likely result of chronic stimulation is increased DNA synthesis, leading to cellular growth and division [1] as may occur during renal failure; the mechanism by which this occurs has not been investigated.

The concentration of Ca^{++} at which half-maximal secretion above basal occurs has been called the 'set point'. Thus, the actual amount of PTH secreted at a given calcium concentration is dependent upon the basal level of secretion, upon this set point, and upon the steepness of the curve by which alterations of Ca^{++} levels affect secretion [28] (fig. 2).

In vitro, this curve is quite steep for normal dispersed parathyroid cells, resulting in a large change in the amount of PTH secreted when calcium levels are altered over a narrow range. Brown [28] has shown a very narrow 'set point' of range for dispersed normal parathyroid cells $(0.97-1.04 \text{ m}M \text{ Ca})$ and a maximal suppressibility of PTH secretion of only $67-79\%$.

Calcium may also control the intracellular degradation of PTH and thus regulate the quantities of intact hormone available for acute secretion. Chu et al. [29] and Habener et al. [30] have shown inhibition of intracellular degradation of PTH by low calcium levels and stimulation of fragment formation by increasing calcium levels. The increased rate of secretion of PTH upon acute decrease of extracellular calcium is explained by release of stored PTH from secretory vesicles.

The mechanism by which calcium regulates PTH secretion has not been established. The parathyroid cell responds to extracellular levels of calcium through alterations of intracellular calcium levels. Addition of the calcium ionophore, A-23187, to media incubating parathyroid cells produces inhibition of PTH secretion at lower extracellular calcium levels than occurs in the absence of the ionophore [31, 32], thus implicating intracellular calcium levels in the regulation of secretion. The rate of release of PTH during short-term incubation in vivo has been found to increase with increasing levels of cAMP. The secretion of newly synthesized PTH does not appear to be increased under these conditions, but stored PTH is preferentially secreted [15]. The addition of dibutyryl cAMP, phosphodiesterase in-

Table I. Correlation of alterations in cAMP levels and PTH release

Agent	Effect on cAMP levels
Agents which inhibit PTH release	
Calcium	↓
α-Adrenergic	↓
Prostaglandin $F_{2\alpha}$	↓
Somatostatin	? ↓
Nitroprusside	↓
Vitamin D metabolites	?
Agents which stimulate PTH release	
Calcitonin	↑
Secretin	↑
Glucagon	↑
Histamine	↑
Cortisol	? ↑
Growth hormone	↑
Epinephrine	↑
Isoproterenol	↑
Dopamine	↑
Prostaglandin E_2	↑
Cholera toxin	↑
Ethanol	↑

hibitors, or isoproterenol, which increase intracellular cAMP levels in parathyroid cells, leads to increased PTH secretion.

The adenylate cyclase of parathyroid cells has been found to be extraordinarily sensitive to inhibition by calcium [33]. Thus, high calcium levels lower intracellular cAMP levels, with the predicted consequence of lowering secretion. Many other agents which affect PTH secretion do so in a direction consistent with their effect on cAMP levels. Increasing levels lead to increased secretion, decreasing levels inhibit secretion (table I).

What regulates the sensitivity of the cyclase to calcium, or the sensitivity of the secretory process to cAMP is not known. However, by analogy with the mechanism by which some of the other inhibitory agents on the table are thought to lower adenylate cyclase activity [34], it is possible that Ca^{++} works through an 'inhibitory' nucleotide regulatory subunit. Such a mechanism would be consistent with the observation that there are high and low affinity sites which interact with calcium, the number of high affinity sites affected by the presence of GTP [35].

Abnormal Parathyroid Tissue and
Calcium Regulation of PTH Secretion

Primary hyperparathyroidism results most frequently from adenoma of a single parathyroid gland [36]. Hyperplasia of four glands is seen in approximately 10–15% of hyperparathyroidism [37–39]. Despite these anatomic differences, the mechanism responsible for hypercalcemia in these disorders appears similar.

Adenomatous parathyroid tissue and dispersed cells suppress their secretion of PTH more than 50% when exposed to high levels of calcium, but the 'set point' for suppression by calcium varies considerably [28, 39, 40]. Similar observations were made for hyperplastic tissue [39–42].

Although an abnormally high set point for calcium suppression of PTH release may explain the pathophysiology of some hyperparathyroid disorders and may reconcile the finding of inappropriately 'normal' PTH levels (by radioimmunoassay) in patients with hypercalcemia and hyperparathyroidism (fig. 3) other factors may also contribute to or entirely account for the clinical disorder.

Because the secretion of PTH by parathyroid cells cannot be completely suppressed [1–3, 15, 22, 28, 42], a simple increase in parathyroid tissue mass above a critical threshold could theoretically produce hypercalcemia. Increased parathyroid tissue in both hyperplasia and parathyroid adenomas could cause the nonsuppressible component of PTH secretion to exceed this threshold (fig. 3). Indeed, an early study by Gittes and Radde [43] demonstrated that the transplantation of several normal parathyroid glands into a single rat produces hypercalcemia. Similarly, the hyperplasia of normal tissue (secondary hyperparathyroidism) due to the hypocalcemia of chronic renal failure will produce hypercalcemia during the period immediately following renal transplantation [44].

Only cells from parathyroid carcinoma displayed autonomous secretion (failure to diminish PTH release at the highest levels of calcium tested) [42, 45]. Heterogeneity of the response of abnormal parathyroid cells to calcium may relate to the etiology of parathyroid adenomas, which appear to arise from a multicellular origin [46]. Hence, parathyroid adenomas may represent a kind of single-gland hyperplasia. Abnormally high calcium set points, or an increased component of nonsuppressible PTH secretion, or a combination of these factors, may be responsible for the pathophysiology of hyperparathyroidism.

Patients undergoing long-term lithium therapy for treatment of manic-

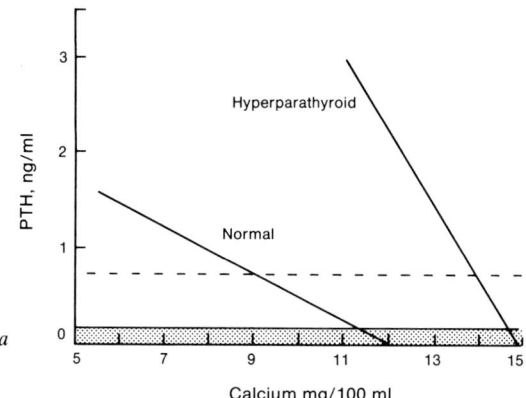

Fig. 3. a Diagram depicting hypothesis of 'set point' error to explain physiological mechanism of abnormality in control of hormone secretion in primary hyperparathyroidism. Ordinate represents plasma parathyroid hormone concentration, and abscissa, plasma calcium.
Hyperparathyroid individuals show steeper slopes of response, owing to large mass of tissue.
Complete suppression of hormone output occurs at abnormally high calcium level. Dashed line
represents upper limit of normal PTH concentration. PTH values within shaded area are below
limits of assay detection.

depressive illnesses have been found to have lower amounts of PTH released
at a given calcium concentration [47]. In in vitro culture of parathyroid cells,
this effect may not be observed for short-term Li^+ treatment [48]. Lithium
appears to increase the 'set point' so that a higher concentration of calcium
is needed to inhibit PTH secretion [49]. Lithium is known to lower cAMP
levels [50]. Thus, it is possible that the effects of Li^+ observed here are due
to its inhibition of cyclase, so that at a given Ca^{++} concentration there is less
cyclase and therefore less secretion; other mechanisms may be involved,
however, as Li^+ has dramatic effects on phosphoinositol turnover [51].

Presumably, increased levels of intracellular cAMP or other 'second
messengers' cause phosphorylation by specific protein kinase(s) of pro-
tein(s) that serve(s) a role in the intracellular transport of hormone or dis-
charge of secretory granules. A cAMP-dependent kinase in parathyroid
cells has been demonstrated [52]. The ratio of basal to cAMP-stimulated
kinase activity was stimulated by secretagogues, but not inhibited by Ca^{++}
or Mg^{++}, at least not with the substrate used in the assay (histone F_{2b}), during short incubations.

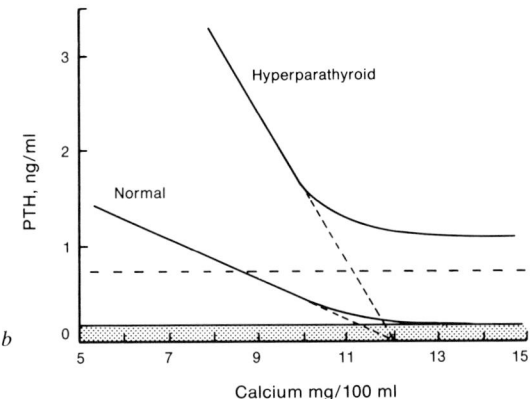

Fig. 3. b Diagram of an alternative hypothesis to explain secretory defect in primary hyperparathyroidism. Hypothesis requires that a small amount of hormone secretion persists despite elevated calcium. In the normal parathyroid, this secretion is small and insignificant. In hyperparathyroidism, greatly increased tissue mass leads to persistent secretion of hormone above normal range (dashed line). Note slopes of both normal and hyperparathyroid extrapolate to same calcium value (12 mg/100 ml). PTH values within shaded area are below limits of assay detection.

However, control of PTH secretion by calcium is probably not entirely mediated by adenylate cyclase and cyclic AMP. When calcium levels are lowered, the secretion of both newly synthesized and stored PTH are stimulated whereas cAMP (or isoproterenol) stimulates the release only of stored PTH [15]. The same is true for release of PSP [17]. What additional mechanism(s) are used by calcium to regulate PTH secretion has not been clearly elucidated. One likely candidate in the regulation of PTH secretion is the calcium/calmodulin complex. Parathyroid glands do contain calmodulin, or a calmodulin-like protein [53]. This protein could serve as a transducer of the calcium concentration signal. For example, calmodulin has been shown to activate a phosphodiesterase in the presence of calcium in another tissue. If it had a similar effect on a phosphodiesterase in the parathyroid glands, it would lower cAMP concentrations as calcium levels were raised. There are cAMP and cGMP phosphodiesterases in parathyroid tissue which are stimulated by the action of calmodulin obtained from another tissue [54].

Calmodulin may also play a role in regulating the intracellular Ca^{++} levels. A calmodulin inhibitor, trifluoperazine, has been found to inhibit a

Ca-ATPase activity in parathyroid cells [55]. Other membrane events, including cell contact [56], hyperpolarization [57], and gradients of other ions [58–60], also have been proposed to affect PTH secretion by altering the intracellular calcium levels.

In a recent preliminary report, it has been observed that phorbol esters (TPA), which are known to activate protein kinase C in other cells, may make PTH secretion insensitive to inhibition at high Ca^{++} concentrations [61]. However, these studies are not conclusive because the concentration of TPA used was in a range known to cause nonspecific membrane fusion in other systems [62].

Other Modulators (Noncalcium) of PTH Secretion

Most studies with other secretagogues for PTH have been conducted in vitro with either collagenase-dispersed parathyroid cells or slices of gland tissue. Since both of these approaches may damage receptors or otherwise damage or alter cells, release a variety of enzymes, or carry along contaminating cells that do not produce PTH, caution must be used in interpreting data from these studies. One attempt to put parathyroid cells in long-term culture resulted in loss of responsiveness to calcium, perhaps due to loss of structured interactions between different types of cells in the parathyroid gland [56]. Thus, it has not been formally established that the same cell which secretes PTH senses the calcium level. In addition, such cells are removed from their normal innervation, which may also result in altered responsiveness to agents such as catecholamines, which may be tonically secreted in close proximity to parathyroid tissue. Hence, although the secretagogues discussed (table I) affect PTH secretion in vitro, their role, if any, in modulation of PTH release in normal physiology is still speculative.

As pointed out in table I, agents which increase cAMP levels in parathyroid cells have been found to increase PTH secretion. (However, it is important to note that increased cAMP levels may not be sufficient stimulus to cause PTH secretion in all tissues [63]. For example, parathyroid cells lacking stored hormone would not be expected to release PTH in response to cAMP since the latter mobilized the stored pool of hormone.) For example, catecholamines, most notably β-adrenergic agents, enhance PTH secretion in vitro [64] and in vivo [65]. The human parathyroid gland is innervated with noradrenergic nerve endings [1, 66] and also responds to circulating catecholamines. Epinephrine and isoproterenol, but not norepinephrine [67], will cause release of PTH, suggesting the importance of β-adrenergic effects over α-adrenergic effects. β-Adrenergic-stimulated secretion of PTH can be blocked in vitro and in vivo by propranolol [68], which may have potential value in treatment of hyperparathyroidism.

β-Adrenergic agonists enhance adenylate cyclase activity [69] and cAMP accumulation [70]. Cyclic AMP increases within 1 min of exposure to β-adrenergic agents and may even reach 60-fold over basal concentration [71]. On the order of 5,000–10,000 β-adrenergic receptors per parathyroid cell have been demonstrated using [^{125}I]iodohydroxybenzylpindolol as radioligand [69].

Dopamine receptors are also present on parathyroid cells. Dopamine increases cAMP levels 20- to 230-fold while increasing PTH releases 2- to 4-fold [72]. At this time, it appears as if catecholamines serve a modulatory rather than regulatory role for PTH secretion. Under conditions of normal or high calcium levels, β-adrenergic agents do not stimulate PTH release [73, 74]. There is also evidence that β-adrenergic agents influence the secretion of only one pool of hormone: that which has been stored intracellularly for more than 60 min [75], consistent with their action via cAMP.

Prostaglandins also can alter release of PTH secretion by parathyroid cells in vitro. Prostaglandin E$_2$ enhances PTH release, which is accompanied by a 2-fold increase in cAMP levels [76]. Prostaglandin F$_{2\alpha}$ inhibits PTH secretion and diminishes cAMP levels [45].

Secretin, glucagon, and histamine may also modulate PTH release from human cells via effects on intracellular levels of cAMP [77, 78].

Other factors that effect PTH release in vitro and may modulate PTH release in vivo include (1) cortisol [79]; (2) growth hormone [80]; (3) somatostatin [81], and (4) ethanol [82] (table I).

In addition to calcium, other ions have been examined for effects on PTH secretion. Disorder of calcium homeostasis accompanies hypomagnesemia, in which PTH levels are low, and target tissues exhibit resistance to PTH. This occurs at very low magnesium concentrations (0.4 mM) and may result from impairment of adenylate cyclase activity [35], although the response of parathyroid cells appears to be more sensitive than other cells to this effect [83]. It has been suggested that under conditions of hypomagnesemia, the adenylate cyclase in parathyroid cells becomes hypersensitive to suppression by calcium [35].

Over a wide concentration range, phosphate ion does not control secretion of PTH [84], although it may have an indirect action, through phosphate effects on calcium levels.

Finally, although tumors rarely [85], if ever, biosynthesize and secrete PTH [86], some tumors may produce a circulating factor which stimulates the parathyroid glands to secrete excess PTH and thus produce hypercalcemia. This mechanism has been postulated as the etiology of hypercalcemia in some cases of leiomyoma [87] and tumors of the plasma cell lineage [88].

Modulation by Other Hormones with Roles in Calcium Homeostasis

Calcitonin in high concentration in vitro stimulates PTH release [89]. In patients with medullary carcinoma of the thyroid, calcium infusion stimulates calcitonin release and thus a paradoxical increase in PTH secretion [90]. Evidence indicating the presence of intracellular receptors for 1,25-dihydroxyvitamin D_3 in parathyroid cells has been obtained [91−93]. In one study, there was an indication that intravenous infusion of 1,25-dihydroxyvitamin D_3 into dialysis patients may shift the set point towards normal, increasing the sensitivity of PTH secretion to calcium [94]. These findings and other preliminary evidence [95] have begun to indicate that there might be a short feedback inhibition by 1,25-dihydroxyvitamin D_3 (the synthesis of which is PTH-stimulated) of PTH synthesis and/or release.

Prolactin, albeit at supraphysiological concentrations, has also been found to stimulate PTH secretion [96]. This could have a physiological role in drawing calcium from the mother's bones during pregnancy and lactation. This effect was reported not to involve alteration in cAMP.

Alterations of PTH Secretion as a Side Effect of Medical Treatment

Lithium, as discussed above, appears to decrease parathyroid cell sensitivity to calcium. Therefore, patients maintained on lithium may have elevated PTH levels and hypercalcemia − a functional hyperparathyroidism − which is reversible with discontinuation of lithium therapy.

Depending upon its concentration, aluminum has been found to suppress or stimulate PTH secretion in vitro [97, 98], but the clinical relevance of this finding is unknown. In cases where aluminum levels are high, such as renal failure, analysis of PTH secretion by radioimmunoassay due to the nature of the clinical disorder is compromised. Furthermore, aluminum accumulates in bone and inhibits bone resorption, which may ultimately lead to a secondary stimulation of PTH secretion. In such cases, the net effect of aluminum on PTH levels in blood is not known.

WR-2721, a radioprotective and chemoprotective agent, inhibits PTH secretion in vitro and in vivo [97]. The mechanism of action of WR-2721 is unknown, but the compound has been postulated to act as a reducing agent within the parathyroid cell, possibly altering the structure of critical cellular components of the secretory apparatus. Direct inhibition of PTH action by WR-2721 at the renal tubular level may also occur.

Cimetidine, an H_2-receptor antagonist appears to lower circulating levels of PTH. However, this decline in PTH is not accompanied by a corresponding decline in calcium concentration, suggesting little, if any, effect on biologically active PTH. Thus, cimetidine may alter PTH degradation within the gland, giving rise to altered secretion of fragments detected by PTH radioimmunoassays. Thus the value of H_2-receptor blockade in the management of hyperparathyroidism is questionable [100, 101].

Conclusion

The investigation of mechanisms and factors responsible for control of PTH secretion permits greater understanding of the regulation of PTH levels in normal physiology and in disorders of calcium metabolism and aids in the search for agents that may ultimately serve a therapeutic role in the treatment of hyperparathyroid disorders.

References

1   Habener, J.F.; Rosenblatt, M.; Potts, J.T., Jr.: Parathyroid hormone: biochemical aspects of biosynthesis, secretion, action, and metabolism. Physiol. Rev. 64: 985–1053 (1984).
2   Cohn, D.V.; Elting, J.: Biosynthesis, processing, and secretion of parathormone and secretory protein-I. Recent Prog. Horm. Res. 39: 181–209 (1983).
3   Kemper, B.: Biosynthesis and secretion of parathyroid hormone; in Cantin, Cell biology of the secretory process, pp. 443–480 (Karger, Basel 1984).
4   Hendy, G.N.; Kronenberg, H.M.; Potts, J.T., Jr.; Rich, A.: Nucleotide sequence of cloned cDNAs encoding human preproparathyroid hormone. Proc. natn. Acad. Sci. USA 78: 7365–7369 (1981).
5   Meyer, D.I.: The signal hypothesis – a working model. TIBS 320 (1982).
6   Walter, P.; Gilmore, R.; Blobel, G.: Protein translocation across the endoplasmic reticulum. Cell 38: 5–8 (1984).

7   Chan, L.; Bradley, W.A.: Signal peptides. Properties and interactions; in Conn, Cellular regulation of secretion and release, p. 301 (Academic Press, New York 1982).

8   Majzoub, J.A.; Rosenblatt, M.; Fennick, B.; Maunus, R.; Kronenberg, H.M.; Potts, J.T., Jr.; Habener, J.F.: Synthetic pre-proparathyroid hormone leader sequence inhibits cell-free processing of placental, parathyroid, and pituitary prehormones. J. biol. Chem. *255:* 11478–11483 (1980).

9   Koren, R.; Burstein, Y.; Soreq, H.: Synthetic leader peptide modulates secretion of proteins from microinjected Xenopus oocytes. Proc. natn. Acad. Sci. USA *80:* 7205–7209 (1983).

10   Briggs, M.S.; Gierasch, L.M.: Exploring the conformational roles of signal sequences: synthesis and conformational analysis of λ-receptor protein wild-type and mutant signal peptides. Biochemistry *23:* 3111–3114 (1984).

11   Kemper, B.; Habener, J.F.; Rich, A.; Potts, J.T., Jr.: Microtubules and the intracellular conversion of proparathyroid hormone to parathyroid hormone. Endocrinology *96:* 903–912 (1975).

12   Rabbani, S.A.; Kremer, R.; Bennett, H.P.J.; Goltzman, D.: Phosphorylation of parathyroid hormone by human and bovine parathyroid glands. J. Biol. Chem. *259:* 2949–2955 (1984).

13   Habener, J.F.; Amherdt, M.; Ravazzola, M.; Orci, L.: Parathyroid hormone biosynthesis. Correlation of conversion of biosynthetic precursors with intracellular protein migration as determined by electron microscope autoradiography. J. Cell Biol. *80:* 715–731 (1979).

14   Habener, J.F.; Maunus, R.; Dee, P.C.; Potts, J.T., Jr.: Early events in the cellular formation of proparathyroid hormone. J. Cell Biol. *85:* 292–298 (1980).

15   Morrissey, J.J.; Cohn, D.V.: Secretion and degradation of parathormone as a function of intracellular maturation of hormone pools. J. Cell Biol. *83:* 521–528 (1979).

16   Hellerman, J.G.; Cone, R.C.; Potts, J.T., Jr.; Rich, A.; Mulligan, R.C.; Kronenberg, H.M.: Secretion of human parathyroid hormone from rat pituitary cells infected with a recombinant retrovirus encoding preproparathyroid hormone. Proc. natn. Acad. Sci. USA *81:* 5340–5344 (1984).

17   Moran, J.R.; Born, W.; Tuchschmid, C.R.; Fischer, J.A.: Calcium-regulated biosynthesis of the parathyroid secretory protein, proparathyroid hormone, and parathyroid hormone in dispersed bovine parathyroid cells. Endocrinology *108:* 2264–2268 (1984).

18   Bhargava, G.; Russell, J.; Sherwood, L.M.: Phosphorylation of parathyroid secretory protein. Proc. natn. Acad. Sci. USA *80:* 878–881 (1983).

19   Morrissey, J.J.; Shofstall, R.E.; Hamilton, J.W.; Cohn, D.V.: Synthesis, intracellular distribution, and secretion of multiple forms of parathyroid secretory protein-I. Proc. natn. Acad. Sci. USA *77:* 6406–6410 (1980).

20   Habener, J.F.; Stevens, T.D.; Tregear, G.W.; Potts, J.T., Jr.: Radioimmunoassay of human proparathyroid hormone. Analysis of hormone content in tissue extracts and in plasma. J. clin. Endocr. Metab. *42:* 520–530 (1976).

21   Benson, R.C., Jr.; Riggs, B.L.; Pickard, B.M.; Arnaud, C.D.: Immunoreactive forms of circulating parathyroid hormone in primary and ectopic hyperparathyroidism. J. clin. Invest. *54:* 175–181 (1974).

22   Morrissey, J.J.; Cohn, D.V.: Regulation of secretion of parathormone and secretory protein-I from separate intracellular pools by calcium, dibutyryl cyclic AMP, and *(1)*-isoproterenol. J. Cell Biol. *82:* 93–102 (1979).

23 Dietel, M.; Dorn-Quint, G.: By-pass secretion of human parathyroid adenomas. A particular intracellular form of rapid adaptation to external stimulation. Lab. Invest. *43:* 116–125 (1980).

24 Keeton, T.K.; Campbell, W.: The pharmacologic alteration of renin release. Pharmac. Rev. *32:* 81–227 (1980).

25 Mayer, G.P.: Parathyroid secretion rate in calves. 55th Annu. Meet. Endocrine Society, 1973, Abstr. 224, p. A-160.

26 Russell, J.; Lettieri, D.; Sherwood, L.M.: Direct regulation by calcium of cytoplasmic messenger ribonucleic acid coding for pre-proparathyroid hormone in isolated bovine parathyroid cells. J. clin. Invest. *72:* 1851–1855 (1983).

27 Heinrich, G.; Habener, J.: Unpubl. observations.

28 Brown, E.M.: Four-parameter model of the sigmoidal relationship between parathyroid hormone release and extracellular calcium concentration in normal and abnormal parathyroid tissue. J. clin. Endocr. Metab. *56:* 572–581 (1983).

29 Chu, L.L.H.; MacGregor, R.R.; Hamilton, J.W.; Cohn, D.V.: Conversion of proparathyroid hormone to parathyroid hormone. The use of amines as specific inhibitors. Endocrinology *95:* 1431–1438 (1974).

30 Habener, J.F.; Kemper, B.; Potts, J.T., Jr.: Calcium-dependent intracellular degradation of parathyroid hormone. A possible mechanism for the regulation of hormone stores. Endocrinology *79:* 431–434 (1975).

31 Habener, J.F.; Stevens, T.D.; Ravazzola, T.; Orci, L.; Potts, J.T., Jr.: Effects of calcium ionophores on the synthesis and release of parathyroid hormone. Endocrinology *101:* 1524–1537 (1977).

32 Brown, E.M.; Gardner, D.G.; Aurbach, G.D.: Effects of the calcium ionophore A23187 on dispersed bovine parathyroid cells. Endocrinology *106:* 133–138 (1980).

33 Matsuzaki, S.; Dumont, J.E.: Effect of calcium ion on horse parathyroid gland adenyl cyclase. Biochim. biophys. Acta *284:* 227–234 (1972).

34 Lefkowitz, R.J.; Caron, M.G.; Stiles, G.L.: Mechanisms of membrane-receptor regulation. Biochemical, physiological, and clinical insights derived from studies of the adrenergic receptors. New Engl. J. Med. *310:* 1570–1579 (1984).

35 Oldham, S.B.; Rude, R.K.; Molloy, C.T.; Lipson, L.G.; Boggs, T.T.: The effects of magnesium on calcium inhibition of parathyroid adenylate cyclase. Endocrinology *115:* 1883–1890 (1984).

36 Habener, J.F.; Potts, J.T., Jr.: Parathyroid physiology and primary hyperparathyroidism; in Avioli, Krane, Metabolic bone disease, vol. 2, pp. 1–147 (Academic Press, New York 1978).

37 Roth, S.I.: Recent advances in parathyroid gland physiology. Am. J. Med. *50:* 612–622 (1971).

38 Wang, C.-A.: Surgery of the parathyroid glands. Adv. Surg. *5:* 109–127 (1971).

39 Wang, C.-A.: The anatomic basis of parathyroid surgery. Ann. Surg. *183:* 271–275 (1976).

40 Habener, J.F.: Responsiveness of neoplastic and hyperplastic parathyroid tissues to calcium in vitro. J. clin. Invest. *62:* 436–450 (1978).

41 Brown, E.M.; Brennan, M.F.; Hurwitz, S.; Windeck, R.; Marx, S.J.; Spiegel, A.M.; Koehler, J.O.; Gardner, D.B.; Aurbach, G.D.: Dispersed cells prepared from human parathyroid glands: distinct calcium sensitivity of adenomas vs. primary hyperplasia. J. clin. Endocr. Metab. *46:* 267–276 (1978).

42 Brown, E.M.; Garner, D.G.; Brennan, M.F.; Marx, S.J.; Spiegel, A.M.; Attie, M.F.; Downs, R.W., Jr.; Doppman, J.L.; Aurbach, G.D.: Calcium-regulated parathyroid hormone release in primary hyperparathyroidism. Am. J. Med. *66:* 923–931 (1979).

43 Gittes, R.R.; Radde, I.C.: Experimental hyperparathyroidism from multiple isologous parathyroid transplants. Homeostatic effect of simultaneous thyroid transplants. Endocrinology *78:* 1015–1022 (1966).

44 Mahaffey, J.; Potts, J.T., Jr.: Secondary hyperparathyroidism: pathophysiology and etiology; in DeGroot, Endocrinology, pp. 739–744 (Grune & Stratton, New York 1979).

45 Brown, E.M.; Gardner, D.G.; Windeck, R.A.; Aurbach, G.D.: Relationship of intracellular 3',5'-adenosine monophosphate accumulation to parathyroid hormone release from dispersed bovine parathyroid cells. Endocrinology *103:* 2323–2333 (1978).

46 Fialkow, P.J.; Jackson, C.E.; Block, M.A.; Greenwald, K.A.: Multicellular origin of parathyroid 'adenomas'. New Engl. J. Med. *297:* 696–698 (1977).

47 Shen, F.-H.; Sherrard, D.J.: Lithium-induced hyperparathyroidism: an alteration of the set-point. Ann. intern. Med. *96:* 63–65 (1982).

48 Spiegel, A.M.; Rudorfer, M.V.; Marx, S.J.; Linnoila, M.: The effect of short term lithium administration on suppressibility of parathyroid hormone secretion by calcium in vivo. J. clin. Endocr. Metab. *59:* 354–357 (1984).

49 Brown, E.M.: Lithium induces abnormal calcium-regulated PTH release in dispersed bovine parathyroid cells. J. clin. Endocr. Metab. *52:* 1046–1048 (1981).

50 Form, J.: Lithium and cyclic AMP; in Johnson, Lithium research and therapy, pp. 485–497 (Academic Press, New York 1975).

51 Berridge, M.J.: Inositol trisphosphate and diacylglycerol as second messengers. Biochem. J. *220:* 345–360 (1984).

52 Brown, E.M.; Thatcher, J.G.: Adenosine 3',5'-monophosphate (cAMP)-dependent protein kinase and the regulation of parathyroid hormone release by divalent cations and agents elevating cellular cAMP in dispersed bovine parathyroid cells. Endocrinology *110:* 1374–1380 (1982).

53 Brown, E.M.; Dawson-Hughes, B.F.; Wilson, R.E.; Adragna, N.: Calmodulin in dispersed human parathyroid cells. J. clin. Endocr. Metab. *53:* 1064–1071 (1981).

54 Willgoss, D.; Jacobi, J.M.; Jersey, J. de; Bartely, P.C.; Lloyd, H.M.: Effect of calcium on cyclic nucleotide phosphodiesterase in parathyroid tissue. Biochem. biophys. Res. Commun. *94:* 763–768 (1980).

55 Dawson-Hughes, B.F.; Underwood, R.H.; Brown, E.M.: Ca-ATPase activity in bovine parathyroid cells. Metabolism *32:* 874–880 (1983).

56 MacGregor, R.R.; Sarras, M.P., Jr.; Houle, A.; Cohn, D.V.: Primary monolayer cell culture of bovine parathyroids: effects of calcium isoproterenol and growth factors. Mol. cell. Endocrinol. *30:* 313–328 (1983).

57 Posillico, J.T.; Leight, G.S., Jr.; Wells, S.A., Jr.; Anderson, N.C., Jr.: Electrical and secretory properties of normal human parathyroid cells: evidence for ionic regulation of PTH secretion; in Cohn, Fujita, Potts, Talmage, Endocrine control of bone and calcium metabolism, pp. 349–352 (Elsevier, Amsterdam 1984).

58 Shoback, D.M.; Brown, E.M.: PTH release stimulated by high extracellular potassium is associated with a decrease in cytosolic calcium in bovine parathyroid cells. Biochem. biophys. Res. Commun. *123:* 684–690 (1984).

59 Gardner, D.G.; Brown, E.M.; Aurbach, G.D.: Inhibition of parathyroid hormone release from dispersed bovine cells by ouabain. Metabolism *32:* 355–358 (1983).

60 Rothstein, M.; Morrissey, J.; Slatopolsky, E.; Klahr, S.: The role of Na^+-Ca^{++} exchange in parathyroid hormone secretion. Endocrinology *111:* 225−230 (1982).

61 Brown, E.M.; Redgrave, J.; Thatcher, J.: Effect of the phorbol ester TPA on PTH secretion. FEBS Lett. *175:* 72−75 (1984).

62 Nishizuka, Y.: The role of protein kinase C in cell surface signal transduction and tumour promotion. Nature, Lond. *308:* 693−697 (1984).

63 Cantley, L.K.; Scott, D.L.; Cooper, C.W.; Mahaffe, D.D.; Leight, G.S.; Thomas, C.G.; Ontjes, D.A.: Divergent effects of forskolin on 3′,5′-cyclic adenosine monophosphate production and parathyroid hormone secretion. Calcif. Tissue Int. *36:* 87−94 (1984).

64 Williams, G.A.; Hargis, G.K.; Bowser, E.N.; Henderson, W.J.; Martinez, N.J.: Evidence for a role of adenosine 3′,5′-monophosphate in parathyroid hormone release. Endocrinology *92:* 687−691 (1973).

65 Fischer, J.A.; Blum, J.W.; Binswanger, U.: Acute parathyroid hormone response to epinephrine in vivo. J. clin. Invest. *52:* 2434−2440 (1973).

66 Norberg, K.A.; Persson, B.; Granberg, P.O.: Adrenergic innervation of the human parathyroid gland. Acta chir. scand. *141:* 319−322 (1975).

67 Sherwood, L.M.; Abe, M.: Adrenergic receptors and the release of parathyroid hormone. J. clin. Invest. *51:* 88a−89a, abstr. 292 (1972).

68 Caro, J.F.; Castro, J.C.; Glennon, J.A.: Effect of long-term propanolol administration on parathyroid hormone and calcium concentration in primary hyperparathyroidism. Ann. intern. Med. *91:* 740−741 (1979).

69 Brown, E.M.; Hurwitz, S.H.; Woodard, C.J.; Aurbach, G.D.: Direct identification of beta-adrenergic receptors on isolated bovine parathyroid cells. Endocrinology *100:* 1703−1709 (1977).

70 Brown, E.M.; Hurwitz, S.H.; Aurbach, G.D.: Beta-adrenergic stimulation of cyclic AMP content and parathyroid hormone release from isolated bovine parathyroid cells. Endocrinology *100:* 1696−1702 (1977).

71 Brown, E.M.; Aurbach, G.D.: Role of cyclic nucleotides in secretory mechanisms and actions of parathyroid hormone and calcitonin. Vitams Horm. *38:* 205−256 (1980).

72 Brown, E.M.; Caroll, R.J.; Aurbach, G.D.: Dopaminergic stimulation of cyclic AMP accumulation and parathyroid hormone release from dispersed bovine parathyroid cells. Proc. natn. Acad. Sci. USA *74:* 4210−4213 (1977).

73 Blum, J.U.; Fischer, J.A.; Hunziker, W.H.; Binswanger, V.; Picotti, G.B.; Da Prada, M.; Guillebeau, A.: Parathyroid hormone responses to catecholamines and to changes in extracellular calcium in cows. J. clin. Invest. *61:* 1113−1122 (1978).

74 Mayer, G.P.; Hurst, J.G.; Barto, J.H.; Keaton, J.A.; Moore, M.P.: Effect of epinephrine on parathyroid hormone secretion in calves. Endocrinology *104:* 1181−1187 (1979).

75 MacGregor, R.R.; Hamilton, J.W.; Cohn, D.F.: The by-pass of tissue hormone stores during the secretion of newly synthesized parathyroid hormone. Endocrinology *97:* 178−188 (1975).

76 Gardner, D.G.; Brown, E.M.; Windeck, R.; Aurbach, G.D.: Prostaglandin E_2 stimulation of adenosine 3′,5′-monophosphate accumulation and parathyroid hormone release in dispersed bovine parathyroid cells. Endocrinology *103:* 577−582 (1978).

77 Windeck, R.; Brown, E.M.; Gardner, D.G.; Aurbach, G.D.: Effect of gastrointestinal hormones on isolated bovine parathyroid cells. Endocrinology *103:* 2020−2026 (1978).

78 Brown, E.M.; Gardner, D.G.; Windeck, R.A.; Hurwitz, S.; Brennan, M.F.; Aurbach,

G.D.: β-Adrenergically stimulated adenosine 3′,5′-monophosphate accumulation in and parathyroid hormone release from dispersed human parathyroid cells. J. clin. Endocr. Metab. *48:* 618−626 (1979).

79   Au, W.Y.W.: Cortisol stimulation of parathyroid hormone secretion. Science *193:* 1015−1017 (1976).

80   Lancer, S.R.; Bowser, E.N.; Hargis, G.K.; Williams, G.A.: The effect of growth hormone on parathyroid function in cats. Endocrinology *98:* 1289−1293 (1976).

81   Hargis, G.K.; Williams, G.A.; Reynolds, W.A.; Chertow, B.B.; Kukreja, S.C.; Bowser, E.N.; Henderson, W.J.: Effect of somatostatin on parathyroid hormone and calcitonin secretion. Endocrinology *102:* 745−750 (1978).

82   Williams, G.A.; Bowser, E.N.; Hargis, G.K.; Kukreja, S.C.; Shah, J.H.; Vora, N.M.; Henderson, W.J.: Effect of ethanol on parathyroid hormone and calcitonin secretion in man. Proc. Soc. exp. Biol. Med. *159:* 187−191 (1978).

83   Cohan, B.W.; Singer, F.R.; Rude, R.K.: End-organ response to adrenocorticotropin, thyrotropin, gonadotropin-releasing hormone, and glucagon in hypocalcemic magnesium deficient patients. J. clin. Endocr. Metab. *54:* 975−979 (1982).

84   Sherwood, L.M.; Mayer, G.P.; Ramberg, C.F.; Kronfeld, D.S.; Aurbach, G.D.; Potts, J.T., Jr.: Regulation of PTH secretion: proportional control by calcium, lack of effect of phosphate. Endocrinology *83:* 1043−1051 (1968).

85   Holtz, G.; Johnson, T.R., Jr.; Schrock, M.E.: Paraneoplastic hypercalcemia in ovarian tumors. Obstet. Gynec., N.Y. *54:* 483−487 (1979).

86   Mundy, G.R.; Ibbotson, K.J.; D'Souza, S.M.; Simpson, E.L.; Jacobs, J.W.; Martin, T.J.: The hypercalcemia of cancer: clinical implications and pathogenetic mechanisms. New Engl. J. Med. *310:* 1718 (1984).

87   Rathaus, M.; Bernheim, J.L.; Griffel, B.; Bernheim, J.; Taragan, R.; Gutman, A.: Hypercalcemie révélatrice d'un leiomyome de l'intestin grêle: presence d'une substance à activité parathormone. Nouv. Presse méd. *8:* 3245−3246 (1979).

88   Stone, N.J.; Duncan, L.A.; Marynick, S.P.; Matthews, J.L.; Sammons, C.A.: Primary hyperparathyroidism in patients with malignant plasma cell dyscrasias; in Cohn, Fujita, Potts, Talmage, Endocrine control of bone and calcium metabolism (Abstract), p. 478 (Elsevier, Amsterdam 1984).

89   Fischer, J.A.; Oldham, S.B.; Sizemore, G.W.; Arnaud, C.D.: Calcitonin stimulation of parathyroid hormone secretion in vivo. Hormone metabol. Res. *3:* 223−228 (1971).

90   Deftos, L.J.; Parthemore, J.G.: Secretion of parathyroid hormone in patients with medullary thyroid carcinoma. J. clin. Invest. *54:* 416−420 (1974).

91   Brumbaugh, P.F.; Hughes, M.R.; Haussler, M.R.; Cytoplasmic and nuclear binding components for $1\alpha,25$-dihydroxyvitamin D_3 chick parathyroid gland. Proc. natn. Acad. Sci. USA *72:* 4871−4875 (1975).

92   Henry, H.L.; Norman, A.W.: Studies on the mechanism of action of calciferol. VII. Localization of 1,25-dihydroxyvitamin D in chick parathyroid glands. Biochem. biophys. Res. Commun. *62:* 781−788 (1975).

93   Hughes, M.R.; Haussler, M.R.: 1,25-Dihydroxyvitamin D_3 receptor in parathyroid glands: preliminary characterization of cytoplasmic and nuclear binding components. J. biol. Chem. *253:* 1065−1073 (1978).

94   Morrissey, J.; Martin, K.; Hruska, K.; Slatopolsky, E.: Abnormalities in parathyroid hormone secretion in primary and secondary hyperparathyroidism. Adv. exp. Biol. Med. *178:* 389−398 (1984).

95   Silver, J.; Russell, J.; Lettieri, D.; Sherwood, L.M.: Vitamin D metabolites suppress cytoplasmic mRNA coding for pre-proparathyroid hormone in isolated parathyroid cells. Clin. Res. *32:* 561A (1984).

96   Magiliola, L.; Forte, L.R.: Prolactin stimulation of parathyroid hormone secretion in bovine parathyroid cells. Am. J. Physiol. *247:* E675–680 (1984).

97   Morrissey, J.; Rothstein, M.; Mayor, G.; Slatopolsky, E.: Suppression of parathyroid hormone secretion by aluminum. Kidney int. *23:* 699–704 (1983).

98   Bourdeau, A.M.; Plachot, J.J.; Cournot, G.; Pointillart, A.; Sachs, C.; Balsan, S.: In vitro effects of aluminum on parathyroid glands: correspondence between hormonal secretion and ultrastructural aspects; in Cohn, Fujita, Potts, Talmage, Endocrine control of bone and calcium metabolism, pp. 230–231 (Elsevier, Amsterdam 1984).

99   Glover, D.; Riley, L.; Carmichael, K.; Spar, B.; Glick, J.; Kligerman, M.M.; Agus, Z.S.; Slatopolsky, E.; Attie, M.; Goldfare, S.: Hypocalcemia and inhibition of parathyroid hormone secretion after administration of WR-2721 (a radioprotective and chemoprotective agent). New Engl. J. Med. *309:* 1137–1141 (1983).

100 Cunningham, J.; Segre, G.; Slatopolsky, E.: Histamine H_2 receptor blockade in patients with hyperparathyroidism. Am. J. Nephrol. *4:* 205–207 (1984).

101 Glaser, B.; Kraiem, Z.; Rotem, M.; Gonda, M.; Bernheim, J.; Sheinfeld, M.: Effect of acute cimetidine administration on indices of parathyroid hormone action in healthy subjects and patients with primary and secondary hyperparathyroidism. J. clin. Endocr. Metab. *59:* 993–997 (1984).

Michael Rosenblatt, MD, Vice President, Biological Research,
Merck Sharp & Dohme Research Laboratories, Division of Merck & Co. Inc.,
West Point, PA 19486 (USA)

Discussion

Kokot: Thank you very much, Dr. Rosenblatt, for your very interesting presentation.

Ritz: Is there just one PTH receptor, or are there several types? The reason why I am asking this question is that there has been work forthcoming from Hermann-Erlee in Leiden, where she was able to stimulate bone resorption with PTH analogues which did not stimulate cyclic AMP.

Rosenblatt: I think the question is a good one. In bone, the cyclic AMP response may not account for all of the actions and metabolic effects of PTH. Cyclic AMP levels rise in seconds, but it takes hours to observe calcium release from bone. I would not be surprised if other second messengers, such as calcium calmodulin or the phosphatidylinositols were involved in expression of PTH biological activity. However, whether there are different classes of PTH receptors I think is very much an open issue. We have photoaffinity labelled receptors in human bone-, canine kidney- and human skin-derived tissue. The receptors appear to be monotonous and have the same molecular weight in every tissue. Functionally, however, for the promotion of vitamin D synthesis by the 1-25-(OH)-α-VitD-hydroxylase, and in some cases for the cyclic AMP response, there may be some divergence between activity of various parathyroid hormone analogues on these effects and on mineral flux. Kenny and his associates have obtained evidence supporting the concept of multiple classes of PTH receptors.

Gross: Has it been shown that the intracellular calcium concentration in parathyroid cells actually varies with extracellular alterations of calcium? Second, is your inhibitor similar to that Slatopolsky has reported, I think the WR-2721, at the last ASN meeting?

Rosenblatt: Let me answer your first question, which is whether extracellular levels of calcium influence intracellular levels in the parathyroid gland cells. The experiments have not been done in a way that exactly addresses your question. If one exposes parathyroid cells to a calcium ionophore, such as A-23187, the inflow of calcium into cells that results will inhibit parathyroid secretion. Calcium channel blockade will wind up promoting PTH secretion. So it does seem as if the intracellular levels of calcium regulate parathyroid secretion. WR-2721 is a radioprotectant. It is a small molecule containing a sulfhydryl group. Its mechanism of action has not been elucidated. It certainly does not act at the PTH receptor. Rather, it appears to inhibit bone resorption and, independently, PTH secretion by the gland. Our antagonist is a polypeptide hormone analogue which is very similar in structure to parathyroid hormone, but is devoid of PTH-like agonist activity. It works by competitively blocking the PTH receptor.

Heidland: I would like to ask two questions: First, is there an ethanol effect on parathyroid gland, a specific or indirect one? I ask this since a decreased serum calcium level may occur in ethanol abuse, which may indicate a decreased parathyroid activity. Second, what is the effect of calcium antagonists in physiologic doses on PTH secretion?

Rosenblatt: In terms of your first question I do not know of a direct ethanol effect on parathyroid secretion. There is some evidence to suggest ethanol effects on bone formation in vitro and calcium release in vivo. In response to your second question calcium antagonists clinically have no effect on parathyroid hormone secretion, but if one takes a parathyroid gland out and puts it in a culture medium, and then adds a calcium antagonist, PTH secretion will be promoted. I think that glands in vitro respond very differently to glands in vivo. They have been freed by collagenase or trypsin and they are also denervated. They are probably not in a setting that reflects normal physiology.

Kurtz: I am very excited by your report because we have set up a cell culture model of juxtaglomerular cells in which we are studying renin release and your statements about the secretion of parathyroid hormone are the same for the renin release, so it seems to be a very similar mechanism. My question is what is the effect of trifluoperazine, the calmodulin antagonist? Do you see an increase in the release of parathyroid hormone?

Rosenblatt: Work by Dawson-Huges, Underwood and Brown indicates that trifluoperazine inhibits the Ca-ATPase in bovine parathyroid glands. This Ca-ATPase extrudes calcium from the intracellular compartment and may regulate cytosolic calcium, thus playing a regulatory role in parathyroid hormone secretion.

Kurtz: What is the relationship between calcium and the activation of adenylate cyclase? Are they independent? That means, does activation of adenylate cyclase have an effect even in the absence of extracellular calcium?

Rosenblatt: Increase in intracellular calcium will lower adenylate cyclase activity and will lower the intracellular levels of cyclic AMP. In the absence of extracellular calcium (by addition of EGTA) both basal and NaF-stimulated cyclase activity are not affected.

Kurtz: This question was because we have the finding that activation of the adenylate cyclase lowers the calcium permeability of the cellular membrane, so it could be vice versa.

Brodde: I have one question about the catecholamine stimulation of PTH release. You have listed, besides other compounds, dopamine in bovine parathyroid gland with receptors for dopamine I (DAI). This is the only physiological example where DAI receptors are working. Do you have any evidence what is in human parathyroid glands? Is dopamine acting via dopamine or other receptors?

Rosenblatt: That is an important question. Unfortunately, we have no data on this subject.

Massry: The parathyroid cell is the only cell in which calcium entry inhibits hormone secretion. Is it possible that this phenomenon is a feedback control? PTH is an ionophore. Now, if PTH raises serum calcium and then by being high it will allow more calcium to get into the gland, if that will continue to stimulate you would have a positive feedback control. The question is, do you know whether PTH by itself regulates its secretion in vitro? In other words, if you have a cell and you stimulate PTH, will it enhance entry of calcium and shut off the gland for PTH production?

Rosenblatt: Brown, I think, has done that experiment in vitro by adding PTH and he has not found any effect, but the same cautions regarding in vitro gland digests apply. I think these in vitro glands may be damaged in the process of bringing them in vitro, so I do not think that your question can be answered conclusively.

Drueke: I did not see on your list aluminum as a possible inhibitor of PTH secretion, although it was exhaustive. As we know, chronic aluminum intoxication not only inhibits the secretion but leads to involution of the gland. Could it not be that aluminum inhibits PTH formation at steps earlier than secretion?

Rosenblatt: It is possible, but the aluminum story is so complicated that I left it out of this overview presentation. Aluminum has been shown to inhibit or stimulate PTH secretion in vitro depending on its concentration. In renal failure aluminum enters bone and inhibits bone resorption. Therefore, it may stimulate parathyroid secretion secondarily. I am just not sure which factors, clinically, will win in the balance.

Kokot: Thank you very much, Dr. Rosenblatt, for an excellent presentation.

Contr. Nephrol., vol. 50, pp. 96–108 (Karger, Basel 1986)

Uremia, Parathyroid Hormone and Carbohydrate Intolerance[1]

Shaul G. Massry[2]

Division of Nephrology and Department of Medicine,
University of Southern California School of Medicine, Los Angeles, Calif., USA

Patients with chronic renal failure display abnormalities in carbohydrate metabolism [1–5]. They almost always have resistance to the peripheral action of insulin [5, 6], while insulin secretion could be normal [4, 7], increased [8, 9] or decreased [3]. Glucose intolerance is, therefore, usually encountered in uremic patients in whom both impaired tissue sensitivity to insulin and impaired secretion of the hormone co-exist [5, 10].

Certain data suggest that parathyroid hormone (PTH) may affect carbohydrate metabolism. Patients with primary hyperparathyroidism may have glucose intolerance [11, 12]. Elevated insulin plasma levels both in the fasting state and in response to glucose [11, 12] as well as insulin resistance [12] have been reported in these patients.

It is plausible, therefore, to suggest that the state of secondary hyperparathyroidism which exists in patients with advanced renal failure [13–16] plays an important role in the genesis of the glucose intolerance of uremia.

Methods

We examined the role of PTH in the glucose intolerance of uremia utilizing intravenous glucose tolerance test (IVGTT) and euglycemic and hyperglycemic clamp studies as described by De Fronzo et al. [17]. The investigations were performed in two groups of dogs with comparable degree and duration of chronic renal failure (CRF) produced by 5/6 nephrectomy; one group (6 dogs) with intact parathyroid glands (NPX) and hence secondary hyperparathyroid-

[1] This work was supported by grant AM 29955 from the National Institute of Arthritis, Diabetes and Digestive and Kidney Diseases.
[2] We would like to thank Ms. Adele Johnson for her secretarial assistance.

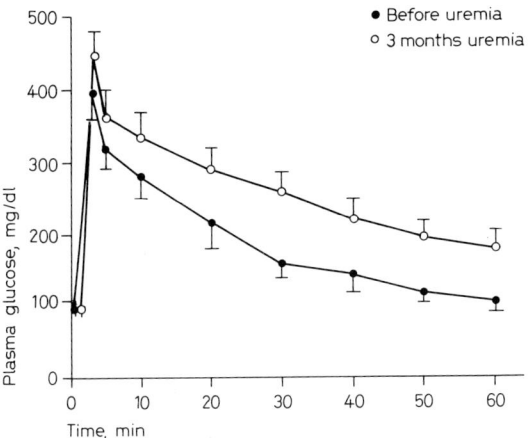

Fig. 1. The changes in plasma glucose concentrations during intravenous glucose tolerance tests performed before (●) and after (○) 3 months of CRF in dogs with intact parathyroid
glands. Each data point represents the mean value of 6 dogs and the brackets denote 1 SE.
Plasma glucose levels in dogs with CRF were significantly higher ($p < 0.01$) at 20–60 min. With
permission from Akmal et al. [18].

ism, and the second group (6 dogs) without the parathyroid glands (NPX-PTX) but maintained
normocalcemic by high intake of calcium. The details of the experimental procedures and the
various techniques and methods utilized in the study were reported elsewhere [18].

Results

The 5/6 nephrectomy resulted in a significant ($p < 0.01$) decrease in
creatinine clearance in both the NPX (from 56 ± 2 to 12 ± 4 ml/min) and the
NPX-PTX (from 58 ± 3 to 13 ± 3 ml/min) dogs and in a significant increase
in serum PTH levels in the NPX animals (from 1.0 ± 0.5 to 37 ± 0.5 pg/ml).
There were no significant differences among the plasma concentrations of
electrolytes before and after the induction of CRF.

The results of the IVGTT before and 3 months after CRF in NPX and
NPX-PTX dogs are shown in figure 1–5. Within 3 min after the injection of
the glucose load, the plasma concentrations of glucose reached their peak
and decreased thereafter. The NPX animals with intact parathyroid glands
and elevated blood levels of PTH displayed glucose intolerance with the
plasma concentrations of glucose being significantly ($p < 0.01$) higher at 20,
30, 40, 50 and 60 min than those observed before CRF (fig. 1). In contrast,

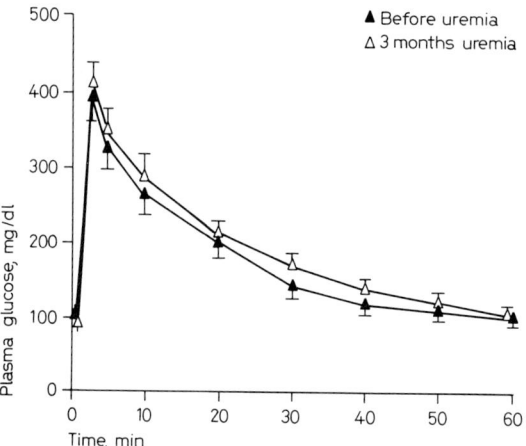

Fig. 2. The changes in plasma glucose concentrations during intravenous glucose tolerance tests performed before (▲) and after (△) 3 months of CRF in parathyroidectomized dogs. Each data point represents the mean value of 6 dogs and the brackets denote 1 SE. There was no significant differences in plasma glucose concentration before and after CRF at all times. With permission from Akmal et al. [18].

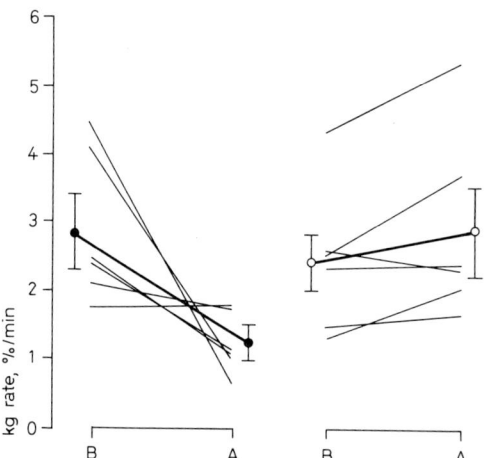

Fig. 3. The kg rate before and 3 months after the CRF in NPX (left panel) and NPX-PTX (right panel). B denotes before CRF and A denotes CRF. Each line represents 1 animal and the heavy lines depict the mean values with brackets denoting 1 SE. The kg rate after CRF was significantly ($p < 0.01$) lower than before CRF in NPX dogs. With permission from Akmal et al. [18].

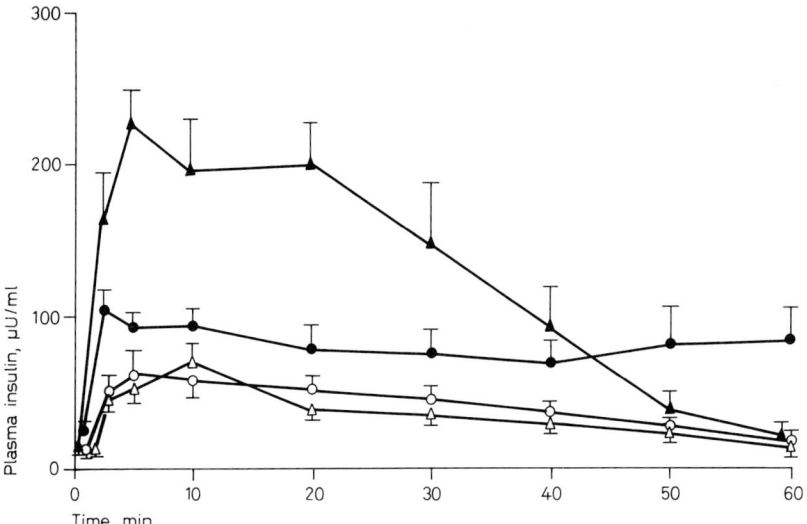

Fig. 4. The changes in plasma insulin concentrations during intravenous glucose tolerance tests performed before (open symbols) and after 3 months of CRF (closed symbols) in NPX (O●) and NPX-PTX (△▲). With permission from Akmal et al. [18].

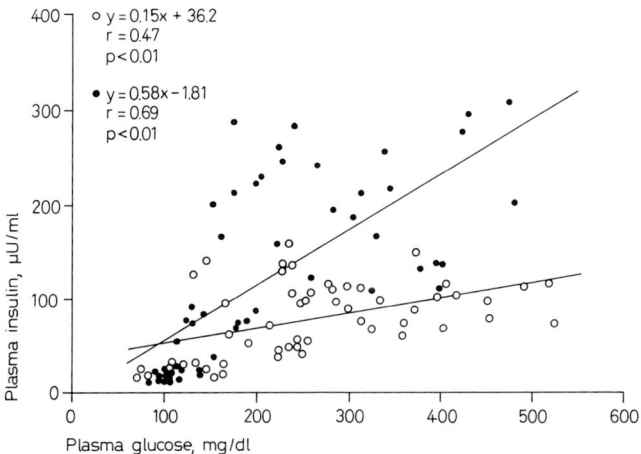

Fig. 5. The relationship between plasma insulin and glucose concentrations observed during intravenous glucose tolerance tests performed in NPX (O) and NPX-PTX (●). With permission from Akmal et al. [18].

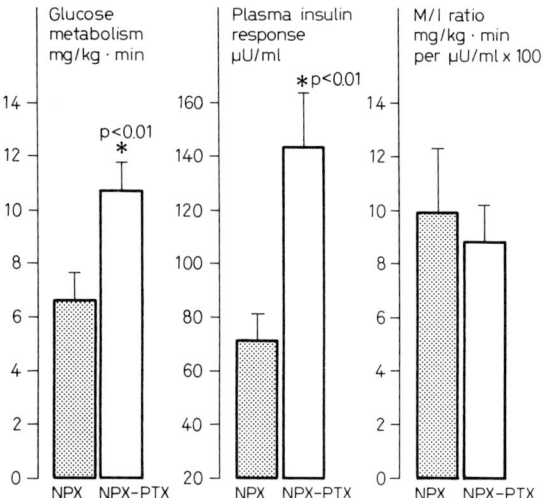

Fig. 6. Glucose metabolism, total insulin response and M/I ratio (total amount of glucose metabolized [M] divided by the total insulin response [I]) observed during the hyperglycemic clamp in NPX and NPX-PTX dogs. Each column represents the mean of data from 6 NPX and 7 NPX-PTX dogs. The brackets denote 1 SE. Asterisks indicate significant difference from NPX with p < 0.01. With permission from Akmal et al. [18].

there was no significant differences between the results of the IVGTT be-fore and after CRF in the NPX-PTX (fig. 2). Thus, these latter animals did not have glucose intolerance.

The K-g rate of glucose (the rate of decline in plasma concentration of glucose) decreased significantly (p < 0.01) after CRF in the NPX dogs (from 2.86 ± 0.48 to 1.23 ± 0.18 %/min) while K-g rate in the NPX-PTX dogs was not affected by CRF (2.41 ± 0.43 vs. 2.86 ± 0.86 %/min, fig. 3).

There were significant increments in plasma insulin levels during IVGTT in all animals. In the NPX dogs, plasma insulin concentrations in-creased from 24 ± 2.3 μU/ml to a peak of 105 mU/ml (p < 0.01) and re-mained elevated throughout the study. In the NPX-PTX dogs, the maximum increment in plasma insulin concentration (from 18 ± 1.2 to 229 ± 19.4 mU/ml) was more than twice that observed in the NPX animals (p < 0.01); the levels gradually declined but were higher than those in the NPX dogs during the first 30 min (p < 0.01) and returned to baseline values by 1 h (fig. 4). These differences in plasma insulin were not due to higher plasma glucose concentrations in NPX-PTX dogs, and for any given level of plasma

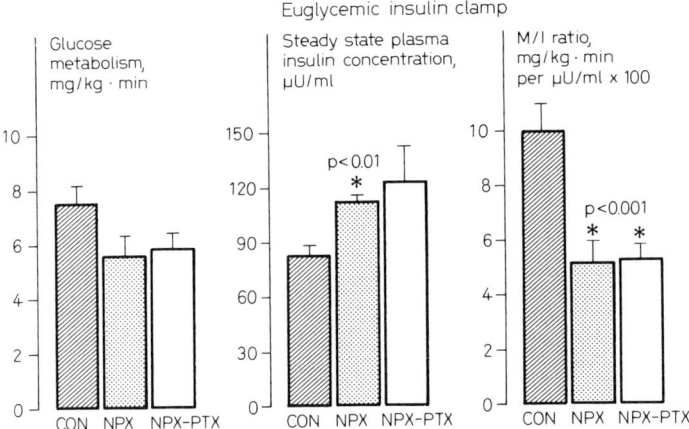

Fig. 7. Glucose metabolism, steady state plasma insulin concentrations and M/I ratio (total amount of glucose metabolized [M] divided by total insulin response [I]) observed during euglycemic insulin clamp in normal dogs and in NPX and NPX-PTX dogs. Each column represents the mean of data from 6 NPX and 7 NPX-PTX dogs. The brackets denote 1 SE. CON = Control. Asterisks denote significant difference (p < 0.01) from control. With permission from Akmal et al. [18].

glucose during IVGTT, the plasma insulin was higher in NPX-PTX than in NPX dogs (fig. 5).

The results of the studies with hyperglycemic clamp are given in figure 6. The total amount of glucose metabolized during the 20 to 120-min period was significantly (p < 0.01) lower by 38% in the NPX compared to the NPX-PTX group (6.64 ± 1.13 vs. 10.74 ± 1.10 mg/kg · min). The early, late and total insulin responses were greater in NPX-PTX animals than in NPX dogs. The total response gave values of 147 ± 31 vs. 72 ± 9 μU/ml (p < 0.025). There was no significant difference between the M/I ratio, a measure of tissue sensitivity to insulin, in the NPX and NPX-PTX dogs (9.9 ± 66 vs. 8.9 ± 1.3 mg/kg · min per μU/ml).

The results of the studies with the euglycemic clamp are presented in figure 7. The total amount of glucose metabolized during elevated blood levels of insulin in the NPX dogs (5.59 ± 0.71 mg/kg · min) was not different from that in the NPX-PTX animals (5.85 ± 0.47 mg/kg · min). Also the M/I ratio was not different among the two groups of dogs (5.12 ± 0.76 vs. 5.18 ± 0.57 mg/kg · min per μU/ml) but both values were significantly (p < 0.01)

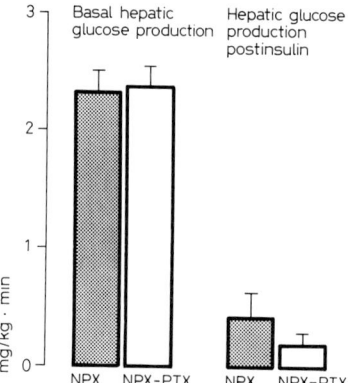

Fig. 8. Basal and postinsulin hepatic glucose production observed during euglycemic insulin clamp in 3 NPX and 3 NPX-PTX dogs. The columns represent the mean data and the brackets denote 1 SE. With permission from Akmal et al. [18].

lower than in normal dogs (9.98 ± 1.26 mg/kg · min per μU/ml). The metabolic clearance rate of insulin was significantly reduced in both NPX (12.1 ± 0.7 mg/kg · min, $p < 0.01$) and NPX-PTX (12.1 ± 0.9 mg/kg · min, $p < 0.02$) dogs as compared with control animals.

Figure 8 provides the results of the studies evaluating basal hepatic glucose production and the response to insulin utilizing the bolus injection and the continuous infusion of [³H]-glucose. Basal hepatic glucose production was similar in NPX and NPX-PTX dogs (2.33 ± 0.32 vs. 2.38 ± 0.35 mg/ kg · min), and the values were not different from those previously reported in normal dogs (2.80 ± 0.20 mg/kg · min) [19].

We also studied the binding affinity, binding sites concentration and binding capacity of monocytes to insulin in the NPX, NPX-PTX and normal dogs. There were no significant differences in these parameters among the three groups of animals.

Discussion

The results of our studies demonstrate that the state of secondary hyperparathyroidism in CRF plays a major role in the genesis of the glucose intolerance in uremia. However, the excess PTH does not affect insulin action on peripheral tissues since the M/I ratio is significantly lower than normal in both NPX and NPX-PTX animals and there is no significant difference in M/I

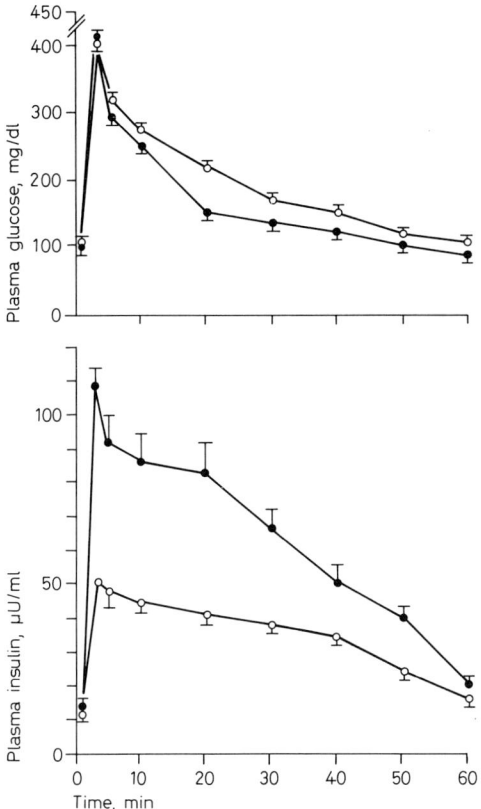

Fig. 9. The changes in plasma glucose and insulin concentrations during intravenous glucose tolerance tests performed in 6 normal dogs (○) and 6 normocalcemic-normophos-
phatemic chronic (6 weeks) thyroparathyroidectomized dogs with normal renal function (●).
Each data point is the mean of 6 animals and the brackets denote 1 SE. Plasma insulin levels
were significantly higher ($p < 0.01$) at all points between 3 and 50 min. With permission from
Akmal et al. [18].

ratio between these two groups of animals. The normalization of the glucose
tolerance in the NPX-PTX dogs must, therefore, be due to improvement in
insulin secretion by the β-cells as both the IVGTT and the clamp studies
demonstrated. It should also be noted that since both the early and late
phases of insulin secretion were enhanced by parathyroidectomy, one must
assume that the release of both the stores as well as the newly synthesized
insulin is enhanced in the NPX-PTX animals. Finally, it should be men-

tioned that the effect of parathyroidectomy on plasma insulin concentration during hyperglycemia is independent of the state of CRF inasmuch as plasma insulin levels during ICGTT in chronic normocalcemic-normophosphatemic-thyroparathyroidectomized dogs were twice (p < 0.01) the levels in normocalcemic dogs with intact parathyroid glands (fig. 9).

In summary, our results indicate that (a) glucose intolerance does not develop with CRF in the absence of PTH; (b) PTH does not affect the metabolic clearance of insulin in CFR, and (c) the normalization of glucose intolerance in CRF in the absence of PTH is due to increased insulin secretion. Thus, the data are consistent with the notion that excess PTH in CRF interferes with the ability of the β-cells to augment insulin secretion appropriately in response to the insulin-resistant state.

The observations of Mak et al. [20] in 8 children before and after medical suppression of the secondary hyperparathyroidism are consistent with our results. They demonstrated that the glucose intolerance in these children disappeared after the normalization of the blood levels of PTH, and the improvement was due to increased insulin response to hyperglycemia after the treatment of the secondary hyperparathyroidism. They concluded that the higher plasma insulin levels overcame the insulin-resistant state, which was not affected by the suppression of the parathyroid gland activity.

During the past several years, data have been accumulated implicating PTH as a uremic toxin [21]. Excess PTH has been shown to exert deleterious effects on central [22−25] and peripheral nervous system, hematopoietic system [28−31], myocardial function [25, 27] and metabolism [32−37], skeletal muscle metabolism [38, 39] and the pressor response to vasoconstrictor agonists [40−42]. The data of the present study demonstrating a role for excess of PTH in the genesis of the carbohydrate intolerance of uremia add new dimensions to the uremic toxicity of PTH.

References

1   Neubauer, E.: Über Hyperglykämie bei Hochdrucknephritis und die Beziehungen zwischen Glykämie und Glucosurie beim Diabetes mellitus. Biochem. Z. *25:* 284−295 (1910).
2   Westervelt, F.B.; Schreiner, G.E.: The carbohydrate intolerance of uremic patients. Ann. intern. Med. *57:* 266−275 (1962).
3   Hampers, C.L.; Soeldoner, J.S.; Doak, P.B.; Merrill, J.P.: Effect of chronic renal failure and hemodialysis on carbohydrate metabolism. J. clin. Invest. *45:* 1719−1731 (1966).
4   Horton, E.S.; Johnson, C.; Lebovitz, H.E.: Carbohydrate metabolism in uremia. Ann. intern. Med. *68:* 63−74 (1968).

5   De Fronzo, R.A.; Andres, R.; Edgar, P.; Walker, W.G.: Carbohydrate metabolism in
     uremia. A review. Medicine 52: 469–481 (1973).
6   De Fronzo, R.A.; Alverstrand, A.; Smith, D.; Hendler, R.; Hendler, E.; Wahren, J.:
     Insulin resistance in uremia. J. clin. Invest. 67: 563–568 (1981).
7   Samaan, N.A.; Freeman, R.M.: Growth hormone levels in severe renal failure. Metabo-
     lism 19: 102–113 (1970).
8   Lowrie, E.G.; Soeldner, J.S.; Hampers, C.L.; Merrill, J.P.: Glucose metabolism and in-
     sulin secretion in uremic, prediabetic, and normal subjects. J. Lab. clin. Med. 76:
     603–615 (1970).
9   Hutching, R.H.; Hagstrom, R.M.; Scribner, B.H.: Glucose intolerance in patients on
     long-term intermittent dialysis. Ann. intern. Med. 65: 275–285 (1966).
10   De Fronzo, R.A.: Pathogenesis of glucose intolerance in uremia. Metabolism 27:
     1866–1880 (1978).
11   Ginsberg, H.; Olefsky, J.M.; Reaven, G.M.: Evaluation of insulin resistance in patients
     with primary hyperparathyroidism. Proc. exp. Biol. Med. 148: 942–945 (1975).
12   Kim, H.; Kalkhoff, R.K.; Costrini, N.V.; Cerletty, J.M.; Jacobson, M.: Plasma insulin
     disturbances in primary hyperparathyroidism. J. clin. Invest. 50: 2596–2605 (1971).
13   Pappenheimer, A.M.; Wilens, S.L.: Enlargement of the parathyroid glands in renal dis-
     ease. Am. J. Path. 11: 73–91 (1935).
14   Roth, S.I.; Marshall, R.B.: Pathology and ultrastructure of the human parathyroid
     glands in chronic renal failure. Archs intern. Med. 124: 390–407 (1969).
15   Berson, S.A.; Yalow, R.: Parathyroid hormone in plasma in adenomatous hyper-
     parathyroidism, uremia, and bronchogenic carcinoma. Science 154: 907–909 (1966).
16   Massry, S.G; Coburn, J.W.; Peacock, M.; Kleeman, C.R.: Turnover of endogenous
     parathyroid hormone in uremic patients and those undergoing hemodialysis. Trans. Am.
     Soc. artif. internal Organs 8: 426–422 (1972).
17   De Fronzo, R.A.; Tobin, J.D.; Andres, R.: Glucose clamp technique. A method for
     quantifying insulin secretion and resistance. Am. J. Physiol. 237: E214–E223 (1979).
18   Akmal, M.; Massry, S.G.; Goldstein, A.D.; Fanti, P.; Weisz, A.; De Fronzo, R.: Role
     of parathyroid hormone in the glucose intolerance of chronic renal failure. J. clin. Invest.
     75: 1037–1044 (1985).
19   Bevilacqua, S.; Barrett, E.; Farranini, E.; Gusberg, R.; Stewart, A.; Richardson, L.;
     Smith, D.; De Fronzo, R.: Lack of effect of parathyroid hormone on hepatic glucose me-
     tabolism in the dog. Metabolism 30: 469–475 (1981).
20   Mak, R.H.; Turner, C.; Haycock, G.B.; Chantler, C.: Secondary hyperparathyroidism
     and glucose intolerance in children with uremia. Kidney int. 24: S123–S133.
21   Massry, S.G.: The toxic effects of parathyroid hormone in uremia. Semin. Nephrol. 3:
     306–328 (1983).
22   Arieff, A.I.; Massry, S.G.: Calcium metabolism of brain in acute renal failure. J. clin. In-
     vest. 53: 387–392 (1974).
23   Guisado, R.; Arieff, A.I.; Massry, S.G.: Changes in the electroencephalogram in acute
     uremia. J. clin. Invet. 55: 738–745 (1975).
24   Cogan, M.G.; Covey, C.M.; Arieff, A.I.; Wisniewski, A.; Clark, O.H.: Central nervous
     system manifestations of hyperparathyroidism. Am. J. Med. 65: 963–970 (1978).
25   Goldstein, D.A.; Feinstein, E.I.; Chui, L.A.; Pattabhiraman, R.; Massry, S.G.: The re-
     lationship between the abnormalities in electroencephalogram and blood levels of PTH
     in dialysis patients. J. clin. Endocr. Metab. 51: 130–134 (1980).

26 Goldstein, D.A.; Chui, L.A.; Massry, S.G.: Effect of parathyroid hormone and uremia on peripheral nerve calcium and motor nerve conduction velocity. J. clin. Invest. *62:* 88–93 (1978).

27 Avram, M.D.; Feinfeld, D.A.; Huatuco, A.H.: Search for the uremic toxin. Decreased motor nerve conduction velocity and elevated parathyroid hormone in uremia. New Engl. J. Med. *298:* 1000–1004 (1978).

28 Meytes, D.; Bogin, E.; Ma, A.; Dukes, P.P.; Massry, S.G.: Effect of parathyroid hormone on erythropoiesis. J. clin. Invest. *67:* 1263–1269 (1981).

29 Bogin, E.; Massry, S.G.; Levi, J.; Djaldeti, M.; Bristol, G.; Smith, J.: Effect of parathyroid hormone on osmotic fragility of human erythrocyte. J. clin. Invest. *69:* 1017–1025 (1982).

30 Massry, S.G.; Doherty, C.C.; Kimball, P.; Moyer, D.; Brautbar, N.: Effect of intact parathyroid hormone (PTH) and its aminoterminal fragment on human polymorphonuclear leukocyte. Implications in uremia. Proc. Am. Soc. Nephrol. *15:* 12A (1983).

31 Remuzzi, G.; Benigni, A.; Dodesini, P.; Schieppati, A.; Livio, M.; Poletti, E.; Mecca, G.; DeGaetano, G.: Parathyroid hormone inhibits human platelet function. Lancet *ii:* 1321–1324 (1982).

32 Lhoste, F.; Drueke, T.; Larus, S.; Boissier, J.R.: Cardiac interaction between parathyroid hormone, β-adrenoreceptor and verapamil in the guinea pig in vitro. Clin. exp. Pharmacol. Physiol. *7:* 377–385 (1980).

33 Drueke, T.; Fleury, I.; Toure, Y.; DeVernejoul, P.; Fauchet, M.; Lesourd, P.; LePailleur, C.; Crosnier, J.: Effect of parathyroidectomy on left ventricular function in haemodialysis patients. Lancet *i:* 112–114 (1980).

34 Bogin, E.; Massry, S.G.; Harary, I.: Effect of parathyroid hormone on heart cells. J. clin. Invest *67:* 1215–1227 (1981).

35 Kahot, Y.; Klein, K.L.; Kaplan, R.A.; Sanborn, W.G.; Kurokawa, K.: Parathyroid hormone has a positive inotropic action on heart. Endocrinology *109:* 2252–2254 (1981).

36 Baczynski, R.; Massry, S.G.; Kohan, R.; Saglikes, Y.; Brautbar, N.: Effect of parathyroid hormone on myocardial metabolism in the rat. Kidney int. *27:* 718–715 (1985).

37 Bogin, E.; Levi, J.; Harary, I.; Massry, S.G.: Effects of parathyroid hormone on oxidative phosphorylation of heart mitochondria. Mineral Electrolyte Metab. *7:* 151–156 (1982).

38 Brautbar, N.; Baczynski, R.; El Belbessi, S.; Massry, S.G.: Effect of PTH on skeletal muscle. Role of PTH in uremic myopathy. Kidney int. *23:* 212 (1983).

39 Garber, A.J.: Effect of parathyroid hormone on skeletal muscle protein and amino acid metabolism in the rat. J. clin. Invest. *71:* 1806–1821 (1983).

40 Saglikes, Y.; Massry, S.G.; Iseki, K.; Nadler, J.L.; Campese, V.M.: Effect of PTH on blood pressure and response to vasoconstrictor agonists. Am. J. Physiol. *248:* F93–F99 (1985).

41 Collins, J.; Massry, S.G.; Campese, V.M.: Parathyroid hormone and the altered vascular response to norepinephrine in uremia. Am. J. Nephrol. *5:* 110–113 (1985).

42 Iseki, K.; Massry, S.G.; Campese, V.M.: Evidence for a role of PTH in the reduced pressor response to norepinephrine in chronic renal failure. Kidney int. *28:* 11–15 (1985).

Shaul G. Massry, MD, Division of Nephrology, University of Southern California School of Medicine, 2025 Zonal Avenue, Los Angeles, CA 90033 (USA)

Discussion

Kokot: Thank you very much, Dr. Massry, for your excellent presentation. It is now open for discussion.

Heidland: These are very fascinating results, but I have one problem: If we assume that PTH is an ionophore, calcium concentration in pancreas should rise and PTH should increase insulin secretion.

Massry: That is correct, but again, Dr. Heidland, we must recognize there are probably two effects of PTH as an ionophore. There is an acute and there is a chronic effect. Acutely, it may increase insulin release and indeed Dr. Kem studied isolated pancreatic cells, and showed that PTH is stimulating insulin release. This is also true in many other functions. If you add PTH to the myocardial myocytes you stimulate both the contractility and heart rate. If you add it to leukocytes you enhance their motility. However, if you prolong the exposure to PTH and then the cells become loaded with a higher concentration of calcium the heart cells stop beating, the leukocytes stop moving and I assume that the pancreas after chronic loading, probably cannot produce more insulin. In order to answer this question, we are currently conducting a study on pancreatic islet cells to evaluate the acute and chronic effect of PTH exposure on insulin release.

Drueke: My question went into the same direction as that of Dr. Heidland. I wanted to know more specifically how your PTX dogs were maintained. They were normocalcemic, I think. This was only by oral calcium, or was it also by $1,25\text{-}(OH)_2D_3$?

Massry: Oral calcium.

Drueke: So these dogs not only lacked the ionophore, but they also lacked $1,25(OH)_2D_3$ in order to get calcium to the pancreatic β-cell. We now know that $1,25(OH)_2D_3$ does induce secretion.

Massry: Of course, $1,25(OH)_2D_3$ does stimulate secretion but both animals will have less $1,25(OH)_2D_3$. It would be very difficult to incriminate the $1,25(OH)_2D_3$.

Hörl: Lindahl and co-workers published a study 1971 in the *Journal of Endocrinology and Metabolism* and showed normal early insulin levels in dialysis patients without hyperparathyroidism and a highly significant increase of insulin secretion in patients with secondary hyperparathyroidism.

Massry: Dr. Hörl, the literature is full of papers from Sweden, Germany, Europe, whatever you like, about patients with renal failure, and about patients with primary hyperparathyroidism. The people you mention, indeed never measured parathyroid hormone level. The serum calcium and phosphate was variable and they claimed there was secondary hyperparathyroidism based on the presence of some bone resorption. You must also remember that serum calcium and serum phosphate are very critical in regulating insulin secretion. That is why in primary hyperparathyroidism you can find anything you want depending on the level of PTH, the level of calcium, and the level of serum phosphate. I do not think that these clinical observations, which are really anecdotal in nature, can provide the answers. Finally, I think you cannot today evaluate the glucose intolerance in uremia or any other insulin resistance state without carefully doing investigations with clamp studies. So I would not relate myself to those studies very seriously.

Ritz: I have a comment and a question. The comment goes along the lines that Drueke asked you. Your argument is that $1,25(OH)_2D_3$ is not involved. The dogs have the same degree of uremia and the only thing that is different is the parathyroid status. Now, of course, those who had higher parathyroid hormone levels might also have had more stimulated $1,25(OH)_2D_3$

secretion by residual renal tissue and the only way to address this criticism would have been serum 1,25(OH)$_2$D$_3$ measurements.

Massry: We did not measure that but if they had higher levels, you would expect them to produce more insulin, not less insulin. That is why it is difficult for me to accept that.

Ritz: If the explanation that you gave to Dr. Heidland's question is correct, i.e. that the action of PTH is mediated by chronic calcium overload of the β-cells, you should be able to prevent it by calcium antagonists. Have you done this experiment?

Massry: We have as yet not done that.

Ritz: And tissue culture experiments with modification of calcium concentrations in the medium?

Massry: We have not. We are just doing it.

Brodde: May I ask you whether there are any changes of insulin receptors in uremic patients?

Massry: We have measured that in dogs in monocytes and found no difference in the PTX dogs or in the animals with intact parathyroid glands. That is surprising because we found no change in the resistance.

Brodde: Is there anything known in uremic patients?

Massry: There are no differences in the receptors and everyone has postulated that the problem is a postreceptor issue.

Ritz: The evidence is in fact somewhat conflicting. In monocytes there is no change of receptor numbers and it is exclusively a postreceptor defect, but in erythrocytes apparently there is a down-regulation. Several authors reported diminished numbers of receptors. But there of course the question is to what extent findings in erythrocytes can be extrapolated to nucleated cells.

Brodde: It is very difficult to get it from the human being.

Kurtz: There is no doubt about the fact that the secretion of insulin upon a stimulus can be divided into two phases. An early phase which can be imitated by calcium ionophores and the sustained phase which can be produced by activators of the protein kinase. Could it be possible that PTH inhibits anyhow the protein kinase.

Massry: Well, I did show you the data on the slide. We have evaluated with our clamp study that only insulin released during the early and late phase and the total insulin release were all better in the PTX examples.

Kokot: Thank you very much.

Contr. Nephrol., vol. 50, pp. 109–118 (Karger, Basel 1986)

Recent Findings on 1,25(OH)$_2$ Vitamin D$_3$ May Provide New Concepts for Understanding the Pathogenesis of Uremia

Eberhard Ritz, Jürgen Merke

Department of Internal Medicine, University of Heidelberg, Heidelberg, FRG

The classical studies of Liu and Chu [1] and Stanbury and Lumb [2] documented that abnormal calcium metabolism of uremic subjects is not corrected by physiological doses of vitamin D but responds to administration of pharmacological doses of the secosterole. These observations led to the concept of 'vitamin D resistance'. These findings were consistent with the later demonstration by Fraser and Kodicek [3] who were the first to demonstrate that 1-alpha-hydroxylation of 25(OH) vitamin D$_3$ occurred exclusively in renal but not extrarenal tissue. This concept readily explained the absence of the bioactive secosterole metabolite 1,25(OH)$_2$ vitamin D$_3$ in the circulation of anephric patients [4]. One would have expected, then, that uremia is a state of absolute 1,25(OH)$_2$D$_3$ deficiency with the consequent disturbances in the vitamin D-regulated target organs of calcium homeostasis.

However, this simple scheme is no longer consistent with recent observations which document (a) synthesis of 1,25(OH)$_2$ vitamin D$_3$ in extrarenal organs and cells, and (b) actions of 1,25(OH)$_2$D$_3$ on organs other than the classical vitamin D target organs involved in calcium homeostasis, i.e. intestine, bone and kidney.

What Is the Evidence for Extrarenal Synthesis of
$1,25(OH)_2$ Vitamin D_3?

Early studies using radioreceptor assays with limited sensitivity reported no detectable levels of $1,25(OH)_2D_3$ in the plasma of patients with advanced renal failure [4]. Several recent observations [5] have confirmed, however, that some material, reacting in radioligand-binding assays and exerting biological activity in bioassay systems, is present even in anephric patients [6]. It must be admitted that interaction with the $1,25(OH)_2D_3$-specific receptor or antibodies in itself is not sufficient proof to identify the substance and, ultimately, characterization of the material by mass spectrometry will be required. However, there are cogent arguments for the identity of the material with $1,25(OH)_2D_3$. First, the material comigrated with $1,25(OH)_2D_3$ in several solvent systems which should have separated $1,25(OH)_2D_3$ from other metabolites, particularly 19-nor-10-keto-25-hydroxyvitamin D_3 [7] or other metabolites. Second, as one experiment of nature, 2 cases [8, 9] of functionally anephric patients with sarcoidosis have been observed who had hypercalcemia and elevated serum $1,25(OH)_2D_3$ levels. This observation is readily explicable by the recent demonstration [10] of 1-alpha-hydroxylase activity in sarcoid granuloma tissue and, more generally, in activated macrophages [11]. This may explain why hypercalcemia is also noted in uremic patients with other granulomatous diseases, e.g. tuberculosis [12]. Because of the recent demonstration of hypercalcemia and $1,25(OH)_2D_3$ overproduction in a patient with silicone granuloma [13], it is of note that our previous studies showed activation of macrophages by silicone particles in experimental animals [14] and dialyzed patients [15].

It has classically be assumed that renal 1-alpha-hydroxylase is tightly regulated so that circulating $1,25(OH)_2D_3$ levels would be independent of the plasma concentration of the precursor substance $25(OH)D_3$. In the extreme case of vitamin D intoxication, $1,25(OH)_2D_3$ levels remain in the normal range despite excessive concentrations of $25(OH)D_3$ [16]. With the introduction of more sensitive assays it has been recognized that even in patients with normal renal function circulating $1,25(OH)_2D_3$ levels change to some extent in response to variations of $25(OH)D_3$ concentrations [17]. This is apparently more marked in uremic patients in whom an increase of $1,25(OH)_2D_3$ concentration was noted with increases of endogenous [18] or exogenous [6, 19] elevations of $25(OH)D_3$ concentrations − similar to patients in sarcoidosis where $1,25(OH)_2D_3$ concentration strikingly rises in parallel with 25(OH)D$_3$ levels. This finding is of interest because it adds a new aspect to vitamin D therapy of uremic patients and will necessitate to modify, at least in part, conventional interpretation of 'vitamin D resistance' of uremia.

Currently, there is no information to what extent such low but finite concentrations of 1,25(OH)$_2$D$_3$ in the circulation of anephric patients abrogate or at least mitigate the expected consequences of 1,25(OH)$_2$D$_3$ deficiency. Because the dissociation equilibrium constant of 1,25(OH)$_2$D$_3$ receptors is in the concentration range of 10^{-10} M, one must anticipate that residual concentrations of 1,25(OH)$_2$D$_3$ as found in the circulation of anephric patients will occupy 1,25(OH)$_2$D$_3$ receptors in target cells. One unknown variable, which will have to be investigated in more detail in the future, may be the modulation of 1,25(OH)$_2$D$_3$ receptor characteristics in uremia. If homologous receptor regulation were present, one would anticipate up-regulation of 1,25(OH)$_2$D$_3$ receptors in uremia. However, we have recently demonstrated [20] that in chronic renal failure maximal specific binding capacity, i.e. receptor density, is diminished rather than increased in various target cells, e.g. Sertoli cells of the testis and basal cells of the epidermis.

Why Does One Not Regularly See Osteomalacia in the Skeleton of Uremic Patients?

Osteomalacia, i.e. deficient mineralisation of bone matrix, is a hallmark of vitamin D deficiency rickets or osteomalacia. It therefore came as a surprise when Bordier et al. [21] reported that osteomalacia is not consistently found in the skeleton of anephric patients despite the presumed absence of 1,25(OH)$_2$D$_3$ in the circulation. This finding has been confirmed by numerous other authors [22].

Even more surprising, skeletal development of children born to uremic mothers who had conceived while being on hemodialysis, is completely normal [23]. This observation is paralleled by our demonstration [24] that skeletal development is completely normal in fetuses of uremic pregnant rats.

It is therefore of note that biosynthesis of 1,25(OH)$_2$D$_3$ has been shown in bone cells of both chick [25] and humans [26]. The identity of the generated metabolite with 1-alpha-25(OH)$_2$ vitamin D$_3$ has been confirmed by mass spectrometry [19]. Furthermore, 1-alpha-hydroxylase activity has been demonstrated in human decidua [27] and placenta [28]. One must

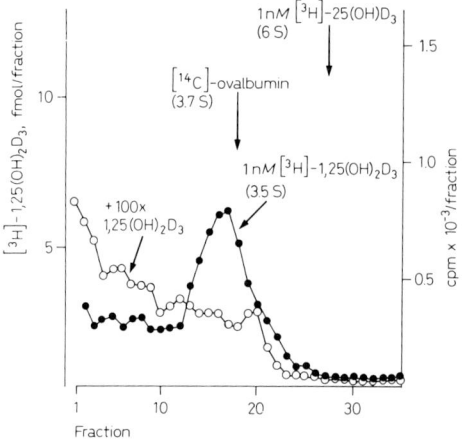

Fig. 1. Sucrose density gradient analysis of the binding of 1,25(OH)₂D₃ and 25(OH)D₃ to the cytosol of basal cells of neonatal mouse epidermis. Cytosol in hypertonic (0.3 *M* KC) KTED buffer (5 × 10⁷ basal cells in 0.2 ml/0.5 mg protein) was incubated for 2 h at 4°C with vitamin D metabolites and sedimented in linear 5–20% (w/v) sucrose density gradients by ultracentrifugation (255,000 *g* for 21 h at 4°C). 1 n*M* [³H]-1,25(OH)₂D₃ in the absence (●—●) or presence (○—○) of 100-fold radioniert 1,25(OH)₂D₃; arrows: [¹⁴C]-ovalbumin marker at 3.7 S, incubation with 1 n*M* [³H]-25(OH)D₃ receptor complex at 3.5 S while the arrow at 6 S with [³H]-25(OH)D₃ confirms the presence of a separate binding site for 25(OH)D₃.

therefore assume that the adult skeleton and the fetal skeleton are exposed to 1,25(OH)₂D₃ by local (possibly paracrine) synthesis of 1,25(OH)₂D₃ in bone and the fetoplacental unit, respectively.

Is 1,25(OH)₂ Vitamin D₃ Involved in Dysfunction in the Uremic Organism of Organs Unrelated to Calcium Homeostasis?

The classical target organs for vitamin D were thought to be intestine, bone and kidneys all of which subserve important roles in the maintenance of constant serum calcium levels and structural integrity of the skeleton.

It has recently been recognized, however, that 1,25(OH)₂D₃ has an important function other than regulating homeostasis of serum calcium. As recently discussed by us in detail elsewhere [29], target organs of 1,25(OH)₂D₃ include, among others, endocrine glands (hypophysis, parathyroids, B cells of the pancreas, ovary and testis), skeletal muscle, epidermis, vascular

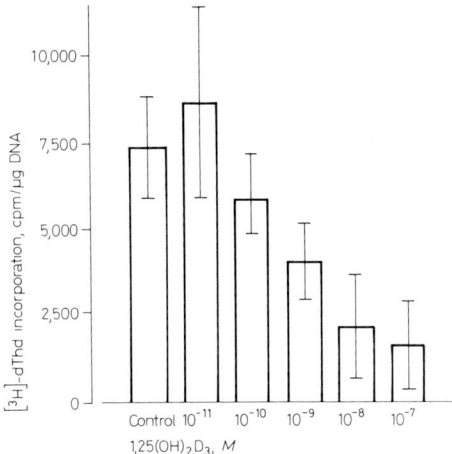

Fig. 2. Effect of increasing concentrations of 1,25(OH)$_2$D$_3$ on growth kinetics (DNA synthesis) of basal keratinocytes. 24 h after seeding of 2×10^6 primary basal cell cultures, medium was replaced by medium containing 0.5% FCS and various concentrations of 1,25(OH)$_2$D$_3$ dissolved in acetone (final concentration of acetone 0.1%). At 24 h cells were labeled for 1 h with 5 µCi/ml medium [^{3}H]-dThd. Radioactivity incorporation in basal cells DNA and DNA content was measured. Values are mean and SD of six incubations. Peak values are significantly different as evaluated by F test (p < 0.001).

smooth muscle, hematolymphatic system and central nervous system. In all these organs, receptors for 1,25(OH)$_2$D$_3$, and actions of 1,25(OH)$_2$D$_3$ on organ function, have been clearly demonstrated.

As one example, figure 1 shows demonstration of receptors for 1,25(OH)$_2$D$_3$ on basal cells of mouse epidermis and figure 2 depicts the effect of 1,25(OH)$_2$D$_3$ on basal epidermal cell proliferation. As illustrated in this cell model, 1,25(OH)$_2$D$_3$ more generally appears to be a signal which decreases cell proliferation and induces cell differentiation and maturation. This has been shown not only in epidermis but also, for instance, in hematolymphatic cells and malignant cell lines.

The multiorgan action of 1,25(OH)$_2$D$_3$ raises the important question whether absolute or relative deficiency of 1,25(OH)$_2$D$_3$ might not be responsible for some aspects of the uremic syndrome [30]. Unfortunately, there is currently no information available in this respect. An easy answer cannot be given, since it is unknown whether low but finite concentrations of 1,25(OH)$_2$D$_3$ in the circulation of uremic patients are sufficient to exert a permissive action, whether local synthesis of $1,25(OH)_2D_3$ in several tissues (e.g. bone, macrophages) might not be sufficient to overcome the lack of the renal endocrine secretion and whether the dose-response relationship for $1,25(OH)_2D_3$ is altered by receptor or postreceptor events in uremia. As a confounding variable, it should be noted that all states of $1,25(OH)_2D_3$ deficiency by necessity are also states of PTH excess. For any organ dysfunction noted in such a condition it must be clearly differentiated to what extent it will be due to PTH excess (which is known to cause multiorgan dysfunction in uremia) [31] and to what extent is it due to the deficiency of $1,25(OH)_2D_3$ itself.

Unfortunately, these considerations must currently remain speculative for want of experimental data. It is hoped, however, that the exciting recent discoveries in vitamin D metabolism will provoke new investigations in this area which will answer the questions raised above.

References

1   Liu, S.H.; Chu, H.I.: Studies of calcium and phosphorus metabolism with special reference to pathogenesis and effects of dihydrotachysterol (A.T 10) and iron. Medicine 22: 103–161 (1943).

2   Stanbury, W.W.; Lumb, G.A.: Metabolic studies of renal osteodystrophy. I. Calcium, phosphorus and nitrogen metabolism in rickets, osteomalacia and hyperparathyroidism complicating chronic uremia and in the osteomalacia of the adult Fanconi syndrome. Medicine, Baltimore 41: 1 (1962).

3   Fraser, D.R.; Kodicek, E.: Unique biosynthesis by kidney of a biologically active vitamin D metabolite. Nature, Lond. 228: 764–771 (1970).

4   Shephard, R.M.; Horst, R.L.; Hamstra, A.J.; DeLuca, H.F.: Determination of vitamin D and its metabolites in plasma from normal and anephric man. Biochem. J. 182: 55–69 (1979).

5   Jongen, M.J.M.; Vijgh, W.J.F. van der; Willems, H.J.J.; Netelenbos, J.C.: Analysis for 1,25-dihydroxyvitamin D in human plasma, after a liquid chromatographic purification procedure, with a modified competitive protein binding assay. Clin. Chem. 27: 444–450 (1981).

6   Lambert, P.W.; Stern, P.H.; Avioli, R.C.; Brackett, N.C.; Turner, R.T.; Greene, A.; Fu, I.Y.; Bell, N.H.: Evidence for extrarenal production of 1,25-dihydroxyvitamin D in man. J. clin. Invest. 69: 722–725 (1982).

7   Matsui, T.; Nakao, Y.; Fujita, T.; Okabe, T.; Ishizuka, S.; Norman, A.W.: Metabolism of $1,25(OH)D_3$ in human promyelocytic leukemia cells (HL 60) isolation and identification of 19-nor-10-keto-25-OH-D_3. Sixth Workshop on Vitamin D, Merano 1985, A 13.

8   Barbour, G.L.; Coburn, J.W.; Slatopolsky, E.; Norman, A.W.; Horst, R.L.: Hypocalcemia in an anephric patient with sarcoidosis; evidence for extrarenal generation of 1,25-dihydroxyvitamin D. New Engl. J. Med. 305: 440–443 (1981).

9   Lemann, J.; Gray, R.; Korkor, A.: Vitamin D and kidney disease; in Robinson, Ne-
    phrology, vol. II. Proc. IXth Int. Congress Nephrology, Los Angeles 1984, pp. 1305–
    1321 (Springer, Berlin 1984).
10 Mason, R.S.; Frankel, T.; Chan, Y.L.; Lissner, D.; Posen, S.: Vitamin D conversion by
    sarcoid lymph node homogenate. Ann. intern. Med. *100:* 59–61 (1984).
11 Mason, R.S.; Frankel, T.; Chan, Y.L.; Lissner, D.; Posen, S.: Vitamin D conversion by
    sarcoid lymphnode homogenate. Ann. intern. Med. *100:* 59–61 (1984).
12 Felsenfeld, A.J.; Drezner, M.; Llach, F.: Hypercalcemia and elevated 1,25(OH)$_2$D lev-
    els in a chronic dialysis patient with tuberculosis. Sixth Workshop on Vitamin D, Merano
    1985, A 434.
13 Kozeny, G.A.; Barbato, A.L.; Bansal, V.K.; Vertano, L.L.; Hano, J.E.: Hypercal-
    cemia associated with silicone-induced granulomas. New Engl. J. Med. *311:* 1103–1105
    (1985).
14 Bommer, J.; Gemsa, D.; Waldherr, R.; Kessler, J.; Ritz, E.: Plastic filing from dialysis
    tubing induces prostanoid release from macrophages. Kidney int. *26:* 331–337 (1984).
15 Bommer, J.; Ritz, E.; Waldherr, R.: Silicone-induced splenomegaly: treatment of pan-
    cytopenia by splenectomy in a dialysis patient. New Engl. J. Med. *305:* 1077–1081
    (1981).
16 Henry, M.: Regulation of the synthesis of 1,25-dihydroxyvitamin D$_3$ and 24,25-dihy-
    droxyvitamin D$_3$ in kidney cell culture; in Kumar, Vitamin D (Martinus Nijhoff, Boston
    1984).
17 Mawer, E.B.; Berry, J.L.; Sommer-Tsilenis, E.; Beykirch, W.; Kuhlwein, A.; Rohde,
    B.T.: Ultraviolet irradiation increases serum 1,25-dihydroxyvitamin D in vitamin D re-
    plete adults. Mineral Electrolyte Metab. *10:* 117–121 (1984).
18 Sorensen, O.H.; Lund, B.I.; Lund, B.J.; Thode, J.D.; Friedberg, M.; Storm, T.; Nis-
    trup, S.: Relationship between circulating 25-OHD and 1,25(OH)$_2$D in end-stage renal
    failure. Evidence for an extrarenal production of 1,25(OH)$_2$D in man. Sixth Workshop
    on Vitamin D, Merano 1985, A 202.
19 Fröhling, P.T.; Schmidt-Gayk, H.; Kokot, F.; Vetter, K.; Mayer, E.; Lindenau, K.: In-
    fluence of vitamin K and keto acids (KA) on 1,25(OH)$_2$D-levels in patients with chronic
    renal failure. Sixth Workshop on Vitamin D, Merano 1985, A 383.
20 Merke, J.; Weber, A.; Hügel, U.; Ritz, E.: Calcitrol receptors in uremia. Proc. Eur.
    Dial. Transplant Ass. *22:* 382–385 (1985).
21 Bordier, P.J.; Tunchot, S.; Eastwood, J.B.; Fournier, A.; Wardener, H.E. de: Lack of
    histological evidence of vitamin D abnormality in bones of anephric patients. Clin. Sci.
    44: 33 (1973).
22 Ritz, E.; Malluche, H.H.; Krempien, B.; Mehls, O.: Bone histology in renal insuffi-
    ciency; in David, Calcium metabolism in renal failure and nephrolithiasis, pp. 197–235
    (Wiley, New York 1977).
23 Confortini, P.; Galanti, G.; Ancona, G.; Giongo, A.; Bruschi, E.; Lorenzini, E.: Full
    term pregnancy and successful delivery in a patient on chronic hemodialysis. Proc. Eur.
    Dial. Transplant Ass. *8:* 74–81 (1971).
24 Ritz, E.; Krempien, B.; Strobel, Z.; Zimmermann, H.; Mehls, O.: Foetal development
    in experimental uraemia. Proc. Eur. Dial. Transplant Ass. *10:* 127–135 (1973).
25 Turner, R.T.; Puzas, E.J.; Forte, M.O.; Lester, G.E.; Gray, T.K.; Howard, G.A.; Bay-
    link, D.J.: In vitro synthesis of 1,25-dihydroxycholecalciferol and 24,25-dihydroxychole-
    calciferol by isolated calvarial cells. Proc. natn. Acad. Sci. USA *77:* 5720–5724 (1980).

26   Howard, G.A.; Turner, R.T.; Sherrard, D.J.; Baylink, D.J.: Human bone cells in cul-
      ture metabolise 25(OH)D$_3$ to 1,25(OH)$_2$D$_3$ and 24,25(OH)D$_3$. J. biol. Chem. *256:*
      7738–7740 (1981).
27   Weisman, Y.; Harell, A.; Edelstein, S.; David, M.; Spirer, Z.; Golander, A.: 1,25-Dihy-
      droxyvitamin D$_3$ and 24,25-dihydroxyvitamin D$_3$ in vitro synthesis by human decidua and
      placenta. Nature, Lond. *281:* 317–319 (1979).
28   Whitsett, J.A.; Ho, M.; Tsang, R.C.; Norman, E.J.; Adams, K.G.: Synthesis of 1,25-
      dihydroxyvitamin D$_3$ in human placenta in vitro. J. clin. Endocr. Metab. *53:* 484–488
      (1981).
29   Merke, J.; Ritz, E.; Boland, R.: Are recent findings on 1,25-Dihydroxycholecalciferol
      metabolism relevant for pathogenesis of uremia? Nephron *42:* 277–284 (1986).
30   Ritz, E.: Pathogenesis of uremia; in Robinson, Nephrology, vol. II. Proc. IXth Int.
      Congr. Nephrology, Los Angeles 1984, pp. 1247–1263.
31   Massry, S.G.: The toxic effects of parathyroid hormone in uremia. Semin. Nephrol. *3:*
      308–330 (1983).

Prof. Dr. Eberhard Ritz, Leiter der Sektion Nephrologie,
Medizinische Universitätsklinik, Rehabilitationszentrum für chronisch Nierenkranke,
Bergheimer Strasse 56a, D–6900 Heidelberg 1 (FRG)

Discussion

Massry· Thank you, Prof. Ritz, for a fascinating and provocative concept. It is open to
discussion. In the slide that you showed in vitamin D deficiency there are so many things, such
as neutrophil impaired motility, and so on. You must remember that any clinical state with vitamin D deficiency is a state of secondary hyperparathyroidism as well. Parathyroid hormone
has been shown in normal conditions without addition of 1,25(OH)$_2$D$_3$ to affect the neutrophil
motility. I think we will need to design some kind of protocol to differentiate between the state
of deficiency and the state of 1,25(OH)$_2$D$_3$ and excess PTH which can occur concomitantly. I
think the documentation of the receptors does not always necessarily mean that receptors have
a function. I think we should not get carried away by the receptor fashion which dominates our
thinking today. As I mentioned before, Sutherland discovered cyclic AMP, it was a great discovery but, at the same time, it hindered progress because we were not ready to think but in
terms of cyclic AMP until our minds opened to other possibilities of action. I think the fashion
of receptors is a terrific concept, but we should have an open mind, that the presence of receptors does not necessarily mean that there is a function and the lack of a receptor does not mean
there will not be an action.

Ritz: I agree with everything you said. Certainly we will have to dissect out the relative
contribution of parathyroid excess and 1,25(OH)$_2$D$_3$ deficiency. This is possible and protocols
are available to test this. I hope that 12 months from now I can discuss this with you on a more
sophisticated level based on experimental data. We also agree with you that the mere demonstration of a receptor does by no means permit one to conclude that it is associated with biologically relevant effects. You know the old cynical saying of endocrinologists: grind and bind. It
is very easy to find receptors once you grind various tissues. I think what we have to do is to
document biological actions and that is precisely the reason why I felt it was important to show

you the proliferation data. I think demonstrations of a receptor are meaningful only if one is also able to attribute biological effects to such receptors. I also fully agree with you in that I keep an open mind with respect to the possibility of nonreceptor-mediated events.

Gross: Are you aware of any differentiating tissues in which no 1,25(OH)$_2$D$_3$ receptors have been found?

Ritz: Yes. For instance spermatogenesis.

Gross: My second question – you showed that in anephric patients the level of measurable 1,25(OH)$_2$D$_3$ was about 30% of normal. You suggested that this 1,25(OH)$_2$D$_3$ came from extrarenal sources. Could you explain again why, given this information in vitamin D intoxication, there should then be no correlation between 25 vitamin D$_3$ and 1,25(OH)$_2$D$_3$?

Ritz: I was very careful to state that with some reservations no relation has been demonstrated between 25(OH)D$_3$ and 1,25(OH)$_2$D$_3$ in vitamin D intoxication of normal subjects. I must point out, however, that as the assays to demonstrate 1,25(OH)$_2$D$_3$ and 25(OH)$_2$D$_3$ have become more sensitive and reliable, a relation between the precursor 25(OH)D$_3$ and the product 1,25(OH)$_2$D$_3$ has been found even in normal persons under a variety of circumstances. As one example, Barbarra Mawer from Manchester published a paper last year in *Mineral Electrolyte Metabolism* that during exposure to sunlight in summer months not only do 25(OH)$_2$D$_3$ levels rise, but also levels of 1,25(OH)$_2$D$_3$. This is accompanied by a rise in urinary calcium. So, even under normal circumstances, some mass action resulting from a precursor product relationship prevails although it is dampened to a large extent, so that it will be detected only with highly sensitive techniques. It may well be that what we see in uremia is just an exaggeration of the same process.

Massry: I think not only sarcoid macrophages produce 1,25(OH)$_2$D$_3$. Dr. Adams has shown that normal macrophages produce 1,25(OH)$_2$D$_3$, so it is not unlikely that you find a correlation.

Ritz: Similar findings have also been demonstrated by Jongen in the Netherlands and by two other groups. The basic observation may therefore be correct, although undoubtedly there are problems with his figures.

Drueke: I would like to come back to that intriguing issue that some groups of workers have found 1,25(OH)$_2$D$_3$ circulating levels in anephric patients and others did not. Of course, you could admit that these differences are only due to methodological problems. I cannot believe that and I would give an alternative suggestion. Some uremic patients who are on dialysis are more or less silicone intoxicated, as your group reported, and they formed granulomas in livers, spleens, and so on, the others do not. So it could be possible that patients who have silicone- or PVC-induced granuloma in their tissues would have a higher propensity of forming 1,25(OH)$_2$D$_3$ than others, even when they are anephric.

Ritz: For this very purpose I mentioned that patients with tuberculosis and silicone-induced granuloma produce great amounts of 1,25(OH)$_2$D$_3$. I perfectly agree with you. There is a report by Leman in the *New England Journal of Medicine* of a case of silicone-induced granuloma with hypercalcemia in whom they were able to demonstrate elevated 1,25(OH)$_2$D$_3$ levels.

Massry: I must say that we have tried to study tuberculosis and we have current studies in about 15 active TB and we could not find higher levels of 1,25(OH)$_2$D$_3$, neither could we find increased production by challenging them with vitamin D.

Ritz: I think it must be a rare event and this is based on the observation that, in India where they have hundreds of tuberculosis patients, during an extended study they found only 4 patients with hypercalcemia. So I think it is an interesting model but not a common event.

Massry: I must say it is fascinating because if you look at the literature 10 or 15 years ago, incidents of hypercalcemia after tuberculosis was about 10−20%. In these days, and we have a big population of active TB in Los Angeles, we do not find hypercalcemia in acute tuberculosis. I think something has happened and the battle is changed.

Ritz: Well, with the wisdom of hindsight, it is ironic that in Europe vitamin D in gigantic doses was given for treatment of tuberculosis in the 1930s, particularly for skin lupus. I am aware of a number of reports of severe vitamin D intoxication. It is most likely that much iatrogenic damage was induced in such cases.

Kokot: May I make some suggestions about the incidence of hypercalcemia in the tuberculosis patients? I think that in some of those with hypercalcemia you did not find elevated $1,25(OH)_2D_3$ levels. I think that some of them did not have tuberculosis but pulmonary sarcoidosis. It is sometimes very difficult to differentiate these two pathological states. It is not excluded that in patients in whom hypercalcemia was reported it was really sarcoidosis and not tuberculosis.

Ritz: In the cases I am referring to, they were documented bacteriologically, although in principle your point is certainly valid.

Contr. Nephrol., vol. 50, pp. 119–129 (Karger, Basel 1986)

Role of Hormonal Disturbances in Uremic Growth Failure

O. Mehls, E. Ritz, G. Gilli, U. Heinrich

Departments of Pediatrics and Internal Medicine, University of Heidelberg, Heidelberg, FRG

Growth is one of the most sensitive indicators of disturbed homeostasis in the growing organism. In the genesis of growth failure of children with renal disease, various factors must be considered, e.g. protein energy malnutrition, hormonal disturbances and accumulation of uremic toxins; their effects may be modified by age at appearance of uremia, primary renal disease, duration of chronic renal failure, metabolic acidosis or other factors. The relative contribution of the above factors has not been established to date [1–3].

This contribution will focus on some hormonal disturbances in uremia which may be related to growth failure of uremic children. An appropriate analysis of this issue first requires a discussion of some methodological problems.

How to Evaluate Bone Age and Puberty?

In a small uremic child, growth may be diminished compared with children of the same chronological age, the same height (height age) or the same degree of bone maturation (bone age). For instance, uremic children with obstructive uropathy are retarded to chronological age but their bone age is advanced to height age [unpubl. observation]. If bone continues to mature in a child who fails to grow, growth potential will be lost. Ultimately, even when uremia is conserved by successful transplantation, adult height will be subnormal.

In the past, it was commonly observed that in uremic children puberty set in late and was delayed in course [4]. In contrast with such earlier reports, our recent studies demonstrate that adult bone age is quickly attained once bone age corresponds to 13 years [5]. The mean chronological age when adult bone age was reached was 16.2 years in uremic girls and 18.1 years in uremic boys, i.e. only marginally different from the normal values of Tanner et al. [6]. There may be some delay in a minority of uremic children but whether this represents a 'tail to the right' of a Gaussian distribution curve remains to be established.

The hormonal changes underlying puberty in uremic children have been clarified to some extent. It has been established that prior to puberty total plasma testosterone (T) levels and free T levels are low [7]. Basal serum levels of LH tend to be high in prepubertal boys and girls [7] but it has not been clarified to what extent this reflects increased secretion or delayed catabolism. Upon stimulation with synthetic releasing hormone (LHRH) similar maximal LH values were reached, but the absolute and fractional increment over baseline was significantly smaller in children with chronic renal failure [7]. The FSH response to LHRH was depressed only in boys and not in girls. It is questionable, however, whether the above finding of diminished gonadotroph reserve in the test situation properly reflects hypothalamic/hypophyseal control of gonadotropin release in normal life. Appropriate measurements of pulsatile gonadotropin secretion, specifically measurements of nocturnal bursts, have not been reported. No major abnormality is suggested by our recent observation that bone age progresses rapidly during puberty and that epiphyseal closure is usually not delayed. In summary, at least with the modalities of treatment currently available and in the absence of malnutrition, delayed puberty is no longer a common problem in uremic children.

Renal Osteodystrophy and Growth

It is commonly assumed that renal osteodystrophy is an important cause of growth failure in uremic children. There is no doubt that in advanced renal osteodystrophy longitudinal growth is impossible when growth plate architecture is disturbed secondary to epiphyseal slipping [8].

More commonly, radiographic evidence of renal osteodystrophy will be unmasked by adequate growth. Indeed, evidence of resorption in the metaphyseal spongiosa and corticalis is pronounced only if and when the local rate of modelling is high as a result of rapid bone growth. As a consequence, we [9] and others [10] did not find a negative correlation between X-ray evidence of renal osteodystrophy and growth rate in uremic children.

Growth and Disturbed Vitamin D Metabolism of Uremia

Stunting of growth is a hallmark of vitamin D dependency rickets type I [11]. Accumulation of labelled $1,25(OH)_2D_3$ in the growth apparatus [12], receptors for $1,25(OH)_2D_3$ in the growth cartilage [13] and metabolism of $1,25(OH)_2D_3$ in chondrocytes of epiphyseal cartilage [14] have been demonstrated. Since bioavailability of the active vitamin D metabolite $1,25(OH)_2D_3$ is diminished in uremia, it was logical to assume that deficiency of $1,25(OH)_2D_3$ was an important cause of impaired growth. This hypothesis found support in the observation of Chesney et al. [15] that $1,25(OH)_2D_3$ promoted growth in some uremic children who had failed to respond to high doses of vitamin D_3. This observation found further support in similar anecdotal reports of other authors [16]. In most of these studies it is impossible to separate the action of $1,25(OH)_2D_3$ from a fortwitous pubertal growth spurt. In any case, none of the above studies found normalization of growth by $1,25(OH)_2D_3$ and the growth-promoting effect of $1,25(OH)_2D_3$ in the original communication of Chesney [15] was only transitory since follow-up of his children showed that the growth curves continued to parallel only the third percentile of age-corrected growth charts [17].

Indeed, several well-controlled prospective long-term studies [18] failed to show a consistent effect of $1,25(OH)_2D_3$ or 1-alpha-hydroxy-cholecalciferol on growth, and this is also in good agreement with our own unpublished experience.

In the past, there has been much controversy whether full expression of the action of vitamin D_3 on the growth apparatus requires only $1,25(OH)_2D_3$ or the concomitant presence of other vitamin D metabolites. To further clarify this point, we compared equicalcemic doses of vitamin D_3 and $1,25(OH)_2D_3$ in subtotally nephrectomized uremic rats [3]. As shown in figure 1, both agents improved, but failed to normalize impaired growth, there being no difference between vitamin D_3 and $1,25(OH)_2$ vitamin D_3.

In the past, several anecdotal reports claimed that dramatic longitudinal growth occurred in uremic children in response to administration of cholecalciferol [1]. Since in none of these studies information on vitamin D status or measurements of circulating $1,25(OH)_2D_3$ levels are available, such growth response with catch-up growth might be explained by vitamin D repletion of vitamin D-deficient uremic children. In dialyzed children vitamin D does not promote growth (fig. 2).

On the whole, the modest effect, if any, of $1,25(OH)_2D_3$ on growth in uremic children is in striking contrast to the effect of $1,25(OH)_2D_3$ in

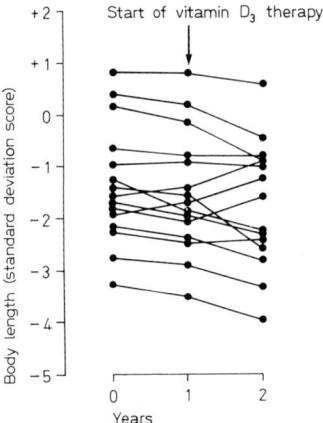

Fig. 1. Growth in dialyzed children. No significant effect of vitamin D therapy.

pseudodeficiency rickets type I, i.e. in selective $1,25(OH)_2D_3$ deficiency, where catch-up growth and normalization of growth rates are observed in response to $1,25(OH)_2D_3$. The findings in uremia would argue against a major role of $1,25(OH)_2D_3$ deficiency for the growth failure in uremia.

Secondary Hyperparathyroidism and Growth

The action of PTH on growth has not been completely clarified. In textbooks, it is commonly stated that children with hypoparathyroidism grow poorly. However, there are no published data in the literature as to whether a growth disturbance occurs in hyperparathyroidism. The separate roles of hypercalcemia and PTH excess are not distinguished to date.

Two clinical observations would even argue for a permissive or even stimulatory role of PTH on growth. When hypoparathyroid children are rendered normocalcemic with vitamin D therapy, they fail to grow normally, possibly because PTH is lacking. On the other hand, children with pseudohypoparathyroidism and osteitis fibrosa are usually tall [19]. Since such children have high circulating PTH levels and since the skeleton (and possibly the growth apparatus) is responsive to PTH despite PTH unresponsiveness of the kidneys, high growth rates and a tall stature may be due to high PTH levels.

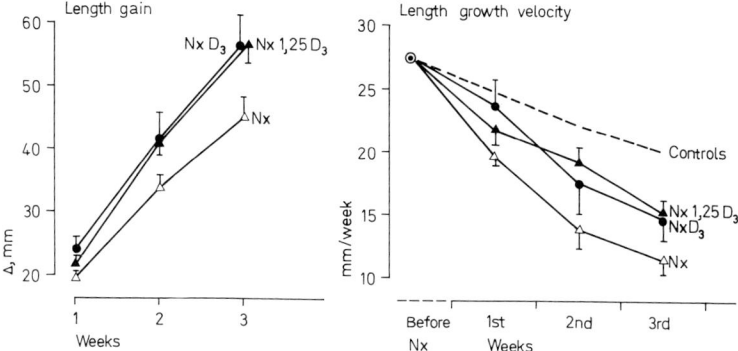

Fig. 2. Growth in experimental uremia. Vitamin D_3 vs. $1,25(OH)_2D_3$. Mean $\pm$ SEM.

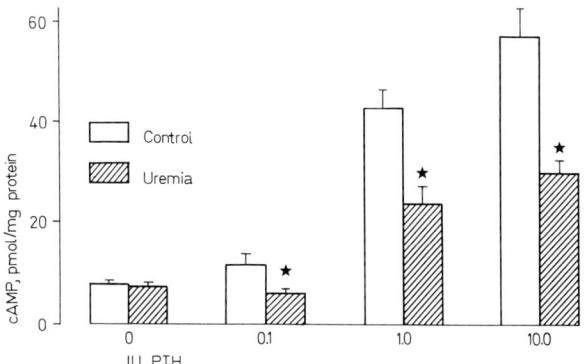

Fig. 3. Influence of PTH on growth cartilage cAMP in vitro. ★ $p < 0,05$.

A stimulatory action of PTH on growth cartilage would parallel similar actions of PTH in osteoprogenitor cells [20] or thymocytes [21] where PTH induces mitosis. It is therefore of note that in experimental uremia we found a diminished response of growth cartilage cAMP to administration of PTH in vivo or addition of PTH in vitro [22], as demonstrated in figure 3.

In the light of the above consideration, it is not surprising that we failed to note a correlation between the severity of X-ray evidence of osteitis fibrosa on the one hand and growth rate on the other. Similarly, Hodson et al. [10] also failed to see a relation between osteitis fibrosa in bone biopsies and growth rate.

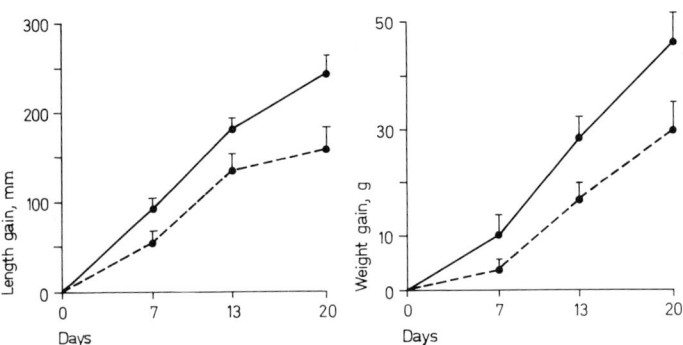

Fig. 4. Growth in experimental uremia. Stimulating effect of growth hormone (GH). ●—● = Uremia + GH; ●‐‐‐● = uremia without GH. Mean ± SEM.

In some children with severest renal osteodystrophy, parathyroidectomy is followed by an increase of growth rate [23–25]. However, in the experience of Broyer [personal commun.] and our own experience, such resumption of growth after parathyroidectomy was only transient and lasted for no more than 7 months. It is conceivable that parallel restoration of the architecture of the growth apparatus, damaged by epiphyseal slipping prior to parathyroidectomy, may underly the above observation.

Growth Hormone and Somatomedin in Uremia

Growth hormone does not directly act on growth cartilage. Its action is mediated by a second hormonal agent, the insulin-like growth factor(s) somatomedin(s), the synthesis and secretion of which in hepatocytes and other tissues is stimulated by growth hormone. Growth hormone levels are clearly elevated in uremia [26, 27], but there is some controversy with respect to levels of somatomedin [= insulin-like growth factors (IGF)]. Bioassays using radiosulfate incorporation into cartilage yield artefactually low concentrations because the ^{35}S tracer is diluted by high levels of sulfate in the uremic plasma [28]. Measurements using radioreceptor or protein-binding assays [29, 30] yielded high somatomedin levels. Recently, Goldberg et al. [31] using a radioimmunoassay, found low IGF I and high IgF II levels associated with a normal protein binding of IGF II.

Table I. Effect of growth hormone (GH) on growth and food conversion to body mass in experimental uremia

	Uremia + GH (n = 9)	Uremia (n = 10)
Cumulative food intake, g	188 ± 28	164 ± 21
Food conversion $\dfrac{\Delta \text{weight, g}}{\text{food intake, g}}$	0.24 ± 0.04	0.17 ± 0.07
Growth increment cm/3 weeks	2.4 ± 0.6	1.6 ± 0.7
g/3 weeks	47 ± 14	30 ± 15

To further evaluate this problem, we compared uremic rats and ad libitum-fed sham-operated control rats. Physiological doses of growth hormone (e.g. doses which stimulate growth in hypophysectomized rats) failed to increase growth rates over that in solvent-treated uremic animals. In contrast, pharmacological doses of growth hormone, i.e. 2.5 U/day by intraperitoneal injection or 0.5–1.0 U/day administered by minipumps, improved growth (fig. 4) and improved the food conversion rate (table I), i.e. the fraction of food (g) which is converted into body mass (g). Improved food conversion was paralleled by lower serum urea levels despite unchanged serum creatinine levels, indicating that growth hormone-treated animals were in an anabolic state. Consequently, we cannot differentiate whether improved growth was due to a direct effect of somatomedin on growth cartilage or due to an indirect effect on anabolism.

IGF carrier protein was significantly higher in growth hormone-treated uremic animals illustrating the well-known stimulatory effect of growth hormone on IGF carrier protein. In contrast, 'total' IGF levels, as measured with the protein-binding assay [32], were unchanged. The finding of increased IGF carrier protein with no change of 'total' IGF serum levels may be due to methodological problems. The IGF fraction comprises both IGF I/somatomedin C which is highly dependent on growth hormone and IGF II which is largely independent of growth hormone. Since the competitive protein-binding assay applied in this study preferentially detects IGF II, an increase of IgF I may have been concealed by this technique.

Table II. Effect of human serum on ³H-thymidine incorporation into growth cartilage (n = 6)

	³H-thymidine activity cpm/mg w/w	p
Uremic serum	3,239 ± 324	<0.01
Control serum	5,403 ± 1,390	

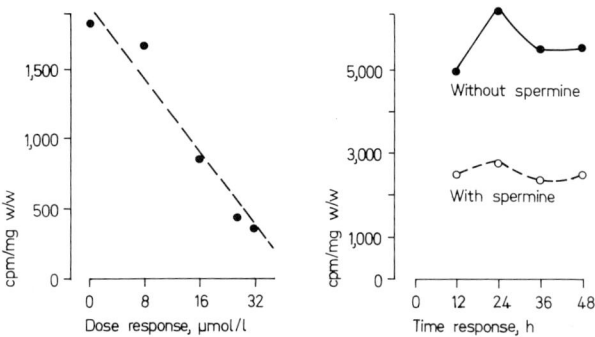

Fig. 5. Inhibitory action of spermine on ³H-thymidine incorporation into growth cartilage.

Blood Spermine and Growth in Uremia

High levels of spermine in uremic plasma have recently been implicated in the genesis of hyporegeneratory anemia of renal failure [33]. Serum of uremic patients inhibited erythroid colony formation in mouse liver cultures and this was reversed by spermine antibodies. This finding raised the possibility that similar inhibition might occur in proliferating growth cartilage. Table II shows that serum of uremic children, but not of control children, inhibits growth cartilage proliferation of the test rats in vitro. Furthermore, spermine caused a dose-dependent inhibition of ³H-thymidine incorporation into growth cartilage [3], as shown in figure 5. However, the relevance of this observation remains to be established, particularly since concentrations chosen are higher than concentrations in predialytic plasma of uremic patients measured by gas chromatography and mass spectrometry [34].

References

1 Mehls, O.; Ritz, E.; Gilli, G.; Kreusser, W.: Growth in renal failure. Nephron *21:* 237 (1978).

2 Schärer, K.; Gilli, G.: Growth in children with renal insufficiency; in Fine, Gruskin, End stage renal disease in children, pp. 271–290 (Saunders, Philadelphia 1984).

3 Mehls, O.; Ritz, E.: Skeletal growth in experimental uremia. Kidney int. *24:* suppl. 15, p. S53 (1983).

4 Schärer, K.; Chantler, C.; Brunner, F.P.; et al.: Combined report on regular dialysis and transplantation of children in Europe 1975. Proc. Eur. Dial. Transplant Ass. *1976:* 60–105.

5 Gilli, G.; Mehls, O.; Wallstein, B.; Schärer, K.: Prediction of adult height in children with chronic renal insufficiency. Kidney int. *24:* suppl. 15, p. S48 (1983).

6 Tanner, J.M.; Whitehouse, R.H.; Takaishi, M.: Standards from birth to maturity of height, weight, height velocity and weight velocity. British children, 1965. Archs Dis. Childh. *41:* 454–471, 613–635 (1966).

7 Oertel, P.J.; Lichtwald, K.; Häfner, S.; Rauh, W.; Schönberg, D.; Schärer, K.: Hypothalamo-pituitary-gonadal axis in children with chronic renal failure. Kidney int. *24:* suppl. 15, p. S34 (1983).

8 Krempien, B.; Mehls, O.; Ritz, E.: Morphological studies on pathogenesis of epiphyseal slipping in uremic children. Virchows Arch. Histol. Path. Anat. *362:* 129 (1974).

9 Ritz, E.; Merke, J.; Mehls, O.: Controversies in vitamin D treatment for renal osteodystrophy; in Norman, Schaefer, Grigoleit, v. Herrath, Vitamin D. Chemical, biochemical and clinical update, p. 1153 (de Gruyter, Berlin 1985).

10 Hodson, E.M.; Shaw, P.F.; Evans, R.A.; Dunstan, C.R.; Hills, E.E.; Wong, S.Y.P.; Rosenberg, A.R.; Roy, L.P.: Growth retardation and renal osteodystrophy in children with chronic renal failure. J. Pediat. *103:* 735 (1983).

11 Prader, A.; Illig, R.; Heierli, E.: Eine besondere Form der primären vitamin D-resistenten Rachitis mit Hypopocalciämie und autosomal dominantem Erbgang. Helv. paediat. Acta *16:* 452 (1961).

12 Dickson, I.; Maher, M.: The influence of vitamin D metabolites on collagen synthesis by chick cartilage in organ culture. J. Endocr. *105:* 79–85 (1985).

13 Chen, T.; Feldmann, D.: Regulation of 1,25-dihydroxyvitamin D_3 receptors in cultured mouse bone cells. J. biol. Chem. *256:* 5561–5566 (1981).

14 Garabedian, M.; Lieberherr, M.; Corvol, M.T.; Balsan, S.: Cellular and subcellular location of the $24,25(OH)_2D_3$ formation in cultured cells from bone and cartilage; in Norman et al., Proc. Fourth Workshop on Vitamin D, Berlin, pp. 391–397 (de Gruyter, Berlin 1979).

15 Chesney, R.W.; Moorthy, A.V.; Eisman, J.A.; Jax, D.K.; Mazzes, R.B.; De Luca, H.F.: Increased growth after long-term oral 1-alpha-25-vitamin D_3 in childhood renal osteodystrophy. New Engl. J. Med. *298:* 238 (1978).

16 Chan, J.C.M.; De Luca, H.F.: Growth velocity in a child on prolonged hemodialysis. Beneficial effect of 1-alpha-hydroxyvitamin D_3. J. Am. med. Ass. *238:* 2053 (1977).

17 Chesney, R.W.: 1,25-Dihydroxy-vitamin D_3 in the treatment of juvenile renal osteodystrophy; in Gruskin, Norman, Pediatric nephrology, p. 209 (Martinus Nijhoff, The Hague 1981).

18 Bulla, M.; Delling, G.; Benz-Bohm, G.; Stock, G.J.; Sánchez de Reutter, A.; Ziegler,

R.; Lühmann, H.; Severin, M.; Kalbitzer, E.; Manegold, C.: Renale Osteodystrophie bei Kindern. Therapieversuch mit 1,25-Dihydroxy-Cholecalciferol. Klin. Wschr. 58: 511–519 (1980).

19 Parfitt, A.M.: Clinical and radiographic manifestations of renal osteodystrophy; in David, Calcium metabolism in renal failure and nephrolithiasis, pp. 145–196 (Wiley, New York 1977).

20 Bingham, P.J.; Barzell, J.A.; Owen, M.: The effect of parathyroid extract on cellular activity and plasma calcium levels in vitro. J. Endocr. 45: 387 (1969).

21 Shelling, D.H.: The parathyroids in health and disease (Mosby, St. Louis 1936).

22 Kreusser, W.; Weinkauf, R.; Mehls, O.; Ritz, E.: Effect of parathyroid hormone, calcitonin and growth hormone on cAMP content of growth cartilage in experimental uremia. Eur. J. clin. Invest. 12: 337–343 (1982).

23 Gilli, G.; Mehls, O.; Ritz, E.: Wirkung von Vitamin D_3 und 1,25$(OH)_2D_3$ auf das Wachstum bei Niereninsuffizienz. Nieren- Hochdruckkrankh. 10: 259 (1981).

24 Broyer, M.; Kleinknecht, C.; Loirat, C.; Marti-Henneberg, C.; Roy, M.P.: Growth in children treated with long-term hemodialysis. J. Pediat. 84: 642 (1974).

25 Mehls, O.: Therapie der urämischen Osteopathie. Mschr. Kinderheilk. 123: 774–776 (1975).

26 Wright, A.D.; Lowy, D.; Fraser, T.R.: Serum growth hormone and glucose intolerance in renal failure. Lancet ii: 798 (1968).

27 Samaan, N.A.; Freeman, R.M.: Growth hormone levels in severe renal failure. Metabolism 19: 102 (1970).

28 Saenger, P.; Wiedemann, E.; Schwartz, E.; Korth-Schmitz, S.; Lewy, J.E.; Riggio, R.R.; Rubin, A.L.; Stenzel, K.H.; New, M.I.: Somatomedin and growth after renal transplantation. Pediat. Res. 8: 163–169 (1974).

29 Schiffrin, A.; Guyda, H.; Robitaille, P.; Posner, B.: Increased plasma somatomedin reactivity in chronic renal failure as determined by acid gel filtration and radio-receptor assay. J. clin. Endocr. Metab. 46: 511 (1978).

30 Takano, K.; Hall, K.; Kastrup, K.W.; Hizuka, N.; Shizume, K.; Kawai, K.; Akimoto, M.; Takuma, T.; Sugino, N.: Serum somatomedin A in chronic renal failure. J. clin. Endocr. Metab. 48: 371–376 (1979).

31 Goldberg, A.C.; Trivedi, B.; Delmez, J.A.; Harter, H.R.; Daughaday, W.H.: Uremia reduces serum insulin-like growth factor I, increases insulin-like growth factor II, and modifies their serum protein binding. J. clin. Endocr. Metab. 55: 1040 (1982).

32 Schalch, D.S.; Heinrich, U.E.; Koch, J.G.; Johnson, C.J.; Schlueter, R.J.: Nonsuppressible insulin-like activity (NSILA). I. Development of a new sensitive competitive protein-binding assay for determination of serum levels. J. clin. Endocr. Metab. 46: 664–671 (1978).

33 Radtke, H.W.; Arvind, B.R.; La Marche, M.B.; Bartos, D.; Bartos, F.; Campbell, R.A.; Fisher, J.W.: Identification of spermine as an inhibitor of erythropoiesis in patients with chronic renal failure. J. clin. Invest. 67: 1623–1629 (1981).

34 Saito, A.; Takagi, T.; Chung, T.G.; Ohta, K.: Serum levels of polyamines in patients with chronic renal failure. Kidney int. 24: suppl. 16, pp. S234–S237 (1983).

Prof. Dr. O. Mehls, Kinderklinik der Universität Heidelberg,
Hofmeisterweg 1–9, D–6900 Heidelberg (FRG)

Discussion

Massry: Dr. Mehls' paper is open for discussion.

Heidland: Can you tell us something about the cortisone levels in children with growth failure? Is there a relationship with enhanced cortisone level and growth retardation?

Mehls: You mean in uremic children? As far as I know, there are no major abnormalities in cortisone levels. There is no correlation between cortisone levels and growth defects in uremia.

Ritz: As far as I am aware, no observation of catch-up growth has been reported in the uremic children treated either conservatively or by transplantation using conventional immunosuppression. However, there have been isolated reports that this occurs with the use of cyclosporin A after transplantation. Has this been experienced by others? There was one report of 2 children from Hannover where this was observed.

Mehls: I think there are anecdotal reports on catch-up growth, in preterminal renal failure without and with vitamin D_3 or $1,25(OH)_2D_3$, in dialyzed children and also in transplanted children. Catch-up growth is mainly reported in very young children under the age of 8 years. This is our experience in Heidelberg, too. I do not think, at the moment, that there is a superior effect of cyclosporin A therapy to conventional therapy. We have to wait for better data.

Massry: Do I understand right that Dr. Chesney's observations are no longer correct?

Mehls: Chesney investigated the promoting effect of $1,25(OH)_2D_3$ for growth. But he showed these effects in only 4 children. Two of these children were in the pubertal stage and possibly showed typical pubertal growth spurts. The others were very small children about the age of 2, where growth rate can change very rapidly. Chesney's follow-up data displayed growth curves only parallel to the third percentile, but not crossing (= catch-up growth). Furthermore, growth curves parallel to the third percentile can be seen without any treatment of vitamin D_3 or $1,25(OH)_2D_3$, in children with chronic renal failure.

Massry: Therefore, we should not agree to $1,25(OH)_2D_3$ for that purpose, or should we?

Mehls: I think you can try it in an individual child, since we have anecdotal positive reports both with vitamin D_3 or with $1,25(OH)_2D_3$. But one should not expect that $1,25(OH)_2D_3$ is superior and solves the growth problems of uremic children.

Dobbelstein: Do you have any data about growth rates in children with nephrotic syndrome and what happens if the glomerulonephritis is healed: Is there any catch-up growth afterwards?

Mehls: Do you mean the nephrotic syndrome with chronic renal failure or without?

Dobbelstein: Without.

Mehls: We analyzed our material recently and it was surprising to see that the vast majority of patients without CRI did not show permanent growth retardation. Many patients who once showed growth retardation had catch-up growth after discontinuing corticosteriod treatment.

Kurtz: Is there information whether the IGF levels increased after dialysis in chronic renal failure?

Mehls: It depends again on the methodology you use. There are conflicting reports. If, for instance, a bioassay technique is used, you will have an increase after dialysis. This result is unreliable because dialysis removes sulfate from the serum which interferes with the assay.

Contr. Nephrol., vol. 50, pp. 130–138 (Karger, Basel 1986)

Gonadal Function in Patients with Acute and Chronic Renal Failure[1]

Franciszek Kokot, Władysław Grzeszczak, Jan Duława

Department of Nephrology, Silesian School of Medicine, Katowice, Poland

Endocrine abnormalities associated with chronic renal failure (CRF) are usually caused by impairment of excretory and biodegrading function of the kidneys, mobilisation of mechanisms counteracting alterations of the internal environment and dietary restrictions and/or medical treatment. Taking into account etiological and pathogenetic differences existing between acute and chronic renal insufficiency, different hormonal abnormalities in these two pathological states may be expected. Among factors which differentiate acute renal failure (ARF) and CRF and simultaneously have a potential influence on endocrine organs the following is to be mentioned: (1) In contrast to chronic renal insufficiency ARF may be caused by different prerenal, renal and postrenal factors which per se may influence the function of endocrine organs. (2) In contrast to CRF, in ARF the amount of functional renal parenchyma and renal blood supply are only moderately reduced [4, 9–11, 26]. (3) Factors initiating ARF are usually not identical with those which are responsible for the maintenance of ARF [1, 12, 19]. As causative and maintaining factors may influence the endocrine system in a different way, hormonal abnormalities may be dependent upon the phase of ARF. (4) In contrast to CRF severity of reduction of (a) renal blood supply; (b) the coefficient of filtration at the glomerulus; (c) glomerular filtration rate; (d) tubular dysfunction and obstruction, and (e) back leakage across the damaged tubule vary from one case of ARF to the other, which in turn may influence the function of the kidney as an excretory and biodegrading organ of hormones. (5) Type of dietary and medical treatment used in pa-

[1] Partially supported by the Polish Ministry of Health and Welfare MZ XIII.

tients with ARF very often differs from that in patients with chronic renal insufficiency. As both nutrition and medicaments may influence function of endocrine organs, different hormonal abnormalities in ARF and CRF may be expected.

From the above follows that endocrine abnormalities in patients with ARF should be related to the etiology, pathogenesis, severity and phase as well as to the kind of medical and dietary treatment of ARF.

The Pituitary Gonadal Axis in ARF

In male patients with ARF during the oliguric/anuric phase [16], basal plasma lutropin (LH) levels are usually significantly elevated. Administration of the lutropin-releasing hormone (LH-RH) to patients with ARF is followed by a normal or even exaggerated and prolonged response of plasma LH during the oliguric phase [16]. During the polyuric phase basal LH levels are still significantly elevated although the LH curve after LH-RH administration is similar to that of normals. Simultaneous administration of LH-RH with naloxone (which is a blocker of opiate receptors) is followed by a significantly higher response of LH secretion than after LH-RH alone.

LH-RH administration to patients with ARF is followed by a normal [16] or suppressed [7, 22] response of plasma FSH levels. In patients with a suppressed response after simultaneous administration of LH-RH and naloxone a normal LH secretion pattern may be observed [7]. Basal plasma FSH concentrations may be significantly elevated [18], normal [16] or even depressed [18] during the anuric/oliguric phase.

In oliguric patients with ARF significantly depressed basal plasma testosterone levels are found [7, 16]. During the polyuric phase plasma testosterone levels are at the lower normal range [16]. During both phases of ARF the response of Leydig cells to LH-RH is blunted [16]. Administration of human chorionic gonadotropin (HCG) to oliguric patients with ARF does not stimulate testosterone secretion [13]. During the polyuric phase a delayed response of plasma testosterone to HCG is found [13, 14].

Moderately elevated basal levels of plasma estradiol are found in male patients during the oliguric phase of ARF [16]. In polyuric patients plasma estradiol concentrations are normal or slightly depressed [16]. Administration of LH-RH is without significant influence on plasma estradiol during both the oliguric and the polyuric phase of ARF [16].

In contrast to healthy subjects, no correlation is found between plasma LH and testosterone levels both in oliguric and polyuric patients with ARF [16]. In oliguric patients with ARF we found a negative correlation between plasma prolactin and testosterone levels, but no significant correlation between plasma parathyroid hormone (PTH) and testosterone concentrations [16]. These last findings are in contrast to those reported by other authors [18, 21].

The pathogenesis of altered LH and FSH levels in ARF is not entirely clear. As the kidneys are an important excretory and biodegrading organ of gonadotropins, impaired renal clearance of these hormones and of LH-RH could contribute to their elevation in blood plasma. This seems unlikely for the following two reasons: (1) In contrast to chronic renal failure in ARF renal blood flow is relatively well preserved [4, 9–11, 25, 26]. This fact suggests that impaired renal clearance of gonadotropins is not the main factor responsible for the stated hormonal abnormalities. (2) Taking into account the biological $T_{1/2}$ of FSH which is about 3–4 h and of LH, which is approximately 50 min, and assuming that impaired renal biodegradation and/or elimination are the main causes of elevated gonadotropin levels, higher FSH levels should be expected during the oliguric than during the polyuric phase of ARF. As already stated, the contrary was found [16]. For these reasons increased LH and FSH secretion rather than diminished renal clearance seems to be responsible for elevated gonadotropin plasma levels in ARF. Significantly elevated LH levels in the presence of low testosterone concentrations suggest that the negative feedback of testosterone on the hypothalamic-pituitary axis is still operating. As no significant correlation was found between plasma LH and testosterone levels [16, 17] existence of both intragonadal and hypothalamic-pituitary defects seem to be likely. Confirmation that the lesion is located among others, within the Leydig cells, is their failure to respond to stimulation by exogenous HCG. As a significant inverse correlation is found between plasma prolactin and testosterone levels [16], participation of enhanced prolactin secretion in the pathogenesis of depressed Leydig cells function seems very likely. It remains to be elucidated whether increased estradiol levels in male patients with ARF are also directly involved in the pathogenesis of hyporesponsiveness of Leydig cells to the stimulatory effect of lutropin. As a positive correlation was found between plasma prolactin and estradiol levels but no correlation between plasma estradiol and testosterone levels [16], it seems that estradiol may depress testicular function indirectly by stimulating prolactin secretion. As responsiveness of both LH and FSH secretion to LH-RH ad-

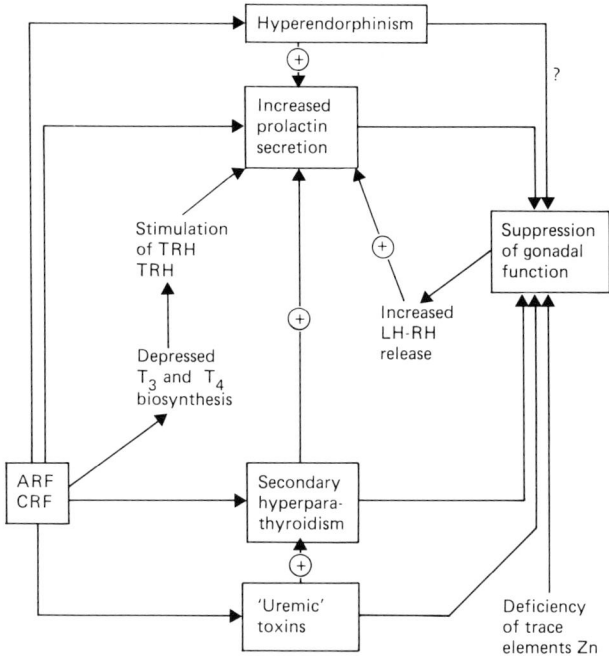

Fig. 1. Relationship of different endocrine abnormalities in the pathogenesis of gonadal hypofunction in uremic patients.

ministration is significantly increased after blockade of opiate receptors by naloxone, while prolactin secretion is significantly suppressed under these conditions, participation of hyperendorphinism in the pathogenesis of gonadal hypofunction in ARF seems very likely [7]. It is to be remembered that under physiological conditions endorphins suppress gonadotropin [23] but stimulate prolactin secretion [24]. In figure 1 a hypothetical scheme of gonadal hypofunction in ARF is presented.

The Pituitary Gonadal Axis in CRF

In male patients with CRF basal plasma LH levels are usually markedly elevated [2, 3, 8, 17, 20, 27–29]. After administration of LH-RH, a subnormal, normal or even exaggerated response of plasma LH [3, 6, 8, 22, 27, 29] may be observed. Simultaneous administration of LH-RH with the opiate receptor blocker naloxone is followed by a significantly higher LH secretion than after LH-RH alone [6].

In female patients with CRF at the reproductive age depressed normal as well as elevated LH levels have been reported. After administration of clomiphene a significant increase of plasma LH may be observed. In contrast, exogenous estrogens do not influence plasma LH levels which suggests the existence of a malfunctioning hypothalamic-pituitary-gonadal feedback [see references in 22].

Basal plasma levels of FSH are most often elevated in male patients with CRF although normal or even depressed concentrations have also been reported [2, 3, 6, 8, 17, 20, 22, 27−29]. After administration of LH-RH or clomiphene, a normal [22] or blunted [6] response of FSH secretion may be found. In hyporesponsive patients administration of naloxone restores a normal FSH secretion pattern to LH-RH [6].

In uremic female patients at the reproductive age rather normal or suppressed basal plasma FSH levels are found with a normal responsiveness to LH-RH or clomiphene respectively [22].

Chronic renal insufficiency is characterized by significantly depressed basal plasma testosterone levels in both sexes [2, 3, 17, 20, 27−29] and blunted or delayed response to HCG [8, 28] or LH-RH [6]. Blocking opiate receptors by naloxone is usually followed by improvement of LH-RH-induced testosterone secretion [6].

From the above, it follows that CRF is characterized by disturbances of gonadotropin and testosterone secretion. Presence of elevated plasma LH and FSH levels seems to be caused not (or not only) by suppressed renal clearance but by enhanced secretion of these hormones. Increased secretion of LH and FSH seems to be evidence of a still-operating hypothalamic-pituitary-gonadal feedback triggered by primary gonadal dysfunction. In turn, the pathogenesis of gonadal dysfunction seems to be complex and multifactorial. Several lines of evidence suggest that the state of secondary hyperparathyroidism, hyperendorphinism, hypersecretion of prolactin, deficiency of trace elements such as zinc, and unidentified uremic toxins may be involved in the pathogenesis of the aboved-mentioned hormonal abnormalities of the hypothalamic-pituitary-gonadal axis in CRF [21] (fig. 1). These hormonal abnormalities together with disturbances in the central and peripheral nervous system (uremic encephalopathy and neuropathy) seem to be critical factors in the pathogenesis of uremic hypogonadism.

From the above, it follows that similar although not identical dysfunction of the hypothalamic-pituitary-gonadal axis may be found both in acute and chronic renal failure. They seem to be due not only to the failing biodegrading and excretory function of the kidneys and compensatory mechanisms secondary to uremia, but also to the altered internal environment, nutritional factors and organic alterations of the endocrine system.

References

1   Börner, H.; Klinkmann, H.: Pathogenesis of acute noninflammatory renal failure. Nephron 25: 262–266 (1980).

2   Czekalski, S.; Malczewska, B.; Sowinski, J.; Sobieszczak, S.; Kozak, W.; Eder, M.; Baczyk, K.: Serum concentration of pituitary, thyroid and gonadal hormone in nondialysed and dialyzed males with chronic renal failure. Proc. Eur. Dial. Transplant Ass. 15: 599–600 (1978).

3   Distiller, L.A.; Morely, J.E.; Sagel, J.; Pokroy, M.; Rabkin, R.: Pituitary-gonadal function in chronic renal failure. The effect of luteinizing releasing hormone and the influence of dialysis. Metabolism 24: 711–720 (1975).

4   Grundfeld, J.P.; Kleinknecht, D.; Moreau, J.F.; Sabato, J.; Kamoun, P.: Hémodynamique intrarénale et secretion de rénine au cours de l'insuffisance rénale aiguë chez l'homme. Resultats préliminaires. J. Urol. Néphrol. 79: 978–983 (1973).

5   Grzeszczak, W.; Kokot, F.; Duława, J.: Influence of naloxone on prolactin secretion in patients with acute and chronic renal failure. Clin. Nephrol. 21: 47–49 (1984).

6   Grzeszczak, W.; Kokot, F.; Duława, J.: Influence of naloxone on lutropin (LH), folitropin (FSH), prolactin and testosterone secretion in patients with chronic renal failure (in press).

7   Grzeszczak, W.; Kokot, F.; Duława, J.: Influence of naloxone on lutropin (LH), folitropin (FSH), prolactin and testosterone secretion in patients with acute renal failure (in press).

8   Holdsworth, S.; Atkins, R.C.; Kretser, D.M. de: The pituitary-testicular axis in men with chronic renal failure. New Eng. J. Med. 296: 1245–1249 (1977).

9   Hollenberg, N.K.; Adams, D.F.; Oken, D.E.; Abrams, H.L.; Merrill, J.P.T.: Acute renal failure due to nephrotoxin. Renal hemodynamic and angiographic studies in man. New Engl. J. Med. 282: 1329–1334 (1970).

10 Hollenberg, N.K.; Epstein, M.; Rosen, S.M.; Basch, R.I.; Oken, D.E.; Merrill, J.P.: Acute renal failure in man. Evidence for preferential renal cortical ischemia. Medicine, Baltimore 47: 455–474 (1968).

11 Hollenberg, N.K.; Sandor, T.; Controy, M.; Adams, D.S.; Solomon, H.S.; Abrams, H.L.; Merrill, J.P.: The transit of radioxenon through the oliguric human kidney. Analysis by the method of maximum likelihood. Kidney int. 3: 177–185 (1973).

12 Kokot, F.: Die Pathophysiologie des akuten, nichtentzündlichen Nierenversagens. Z. Ges. inn. Med. 33: 329–335 (1978).

13 Kokot, F.: The endocrine system in patients with acute renal failure. Proc. Eur. Dial. Transplant Ass. 18: 617–629 (1981).

14 Kokot, F.: Endocrine system in acute renal failure; in Andreucci, Acute renal failure, pathophysiology, prevention and treatment, pp. 167–175 (Nijhoff, Boston 1984).

15 Kokot, F.; Grzeszczak, W.; Duława, J.: Besteht eine Beziehung zwischen der Parathor-

mon- und Prolaktinsekretion bei Kranken mit akutem und chronischem Nierenversagen? Z. Ges. inn. Med. *39:* 40–43 (1984).

16   Kokot, F.; Mleczko, Z.; Pazera, A.: Parathyroid hormone and prolactin and function of the pituitary gonadal axis in male patients with acute renal failure. Kidney int. *21:* 84–89 (1982).

17   Krølner, B.: Serum levels of testosterone and luteinizing hormone in patients with chronic renal disease. Acta med. scand. *205:* 623–627 (1979).

18   Levitan, D.; Moser, S.; Goldstein, D.A.; Kletzky, O.; Massry, S.C.: Disturbances in the function of the hypothalamic-pituitary gonadal (H-P-G) axis during acute renal failure (ARF) in the male (Abstract). Kidney int. *19:* 131 (1981).

19   Levinsky, K.G.: Pathophysiology of acute renal failure. New Engl. J. Med. *296:* 1453–1458 (1977).

20   Lim, V.S.; Fang, W.R.: Gonadal dysfunction in uremic men. Am. J. Med. *58:* 655–662 (1975).

21   Massry, S.G.; Goldstein, D.A.; Procci, W.R.; Kletzky, O.A.: On the pathogenesis of sexual dysfunction of the uraemic male. Proc. Eur. Dial. Transplant Ass. *17:* 139–145 (1980).

22   Osten, B.; Kokot, F.; Klinkmann, H.: Endokrinologische Störungen bei chronischer Niereninsuffizienz und bei Dauerdialysebehandlung. Teil 1. Z. Dt. Gesundh.-Wesen *37:* 2113–2116, 2196–2200 (1982).

23   Quigley, M.E.; Yen, S.S.G.: The role of endogenous opiates on LH secretion during the menstrual cycle. J. clin. Endocr. Metab. *51:* 179–181 (1980).

24   Ragavan, V.V.; Frantz, A.G.: Opioid regulation of prolactin secretion: evidence for a specific role of beta-endorphin. Endocrinology *109:* 1769–1771 (1981).

25   Reubi, F.C.: The pathogenesis of anuria following shock. Kidney int. *5:* 106–110 (1974).

26   Reubi, F.C.; Vorburger, C.: Renal hemodynamics in acute renal failure after shock in man. Kidney int. *10:* 137–143 (1976).

27   Schalch, D.S.; Gonzales-Barcena, D.; Kastin, A.J.: Landa, L.; Lee, L.A.; Zamora, M.T.; Schally, A.V.: Plasma gonadotropin after administration of LH-releasing hormone in patients with renal or hepatic failure. J. clin. Endocr. Metab. *41:* 921–925 (1975).

28   Steward-Bentley, M.; Gans, D.; Horton, R.: Regulation of gonadal function in uremia. Metabolism *23:* 1063–1072 (1974).

29   Swamy, A.P.; Woolf, P.D.; Castero, R.V.M.: Hypothalamic-pituitary-ovarian axis in uremic women. J. clin. Lab. Med. *93:* 1066–1072 (1979).

Prof. Dr. Franciszek Kokot, Silesian School of Medicine, Institute of Internal Medicine, Department of Nephrology, Ul. Francuska 20, 40–027 Katowice (Poland)

Discussion

Ritz: You conclude from your observation that naloxone enhances the response if there is a state of endogenous hyperendorphinism. Is this conclusion really justified? If you pharmacologically intervene and you see an effect, this does not tell that there is an excess of the endogenous substance. Let me give you an example. If you administer converting enzyme inhibitor to a patient with pheochromocytoma, you will see a blood pressure decrease although this does not tell you that primarily you have an increase of renin activity. Might it not be that you are using a pharmacological probe?

Kokot: I can say from experimental data that I think that our conclusions are right. There are some pharmacological data which exclude the direct influence of naloxone on the end organ, i.e. on the endocrine cells themselves. As we observed hormonal changes after blocking opiate receptors, it seems reasonable to accept that hyperendorphinism is important in the regulation of gonadal hormone secretion.

Ritz: You saw the effect in nonuremic individuals as well, didn't you?

Kokot: Yes, of course, but the responsiveness in patients, especially with acute renal failure, is much higher to the administration of naloxone than in normals. This is why we are concluding that, judging from the magnitude of the response, after the administration of naloxone in patients with acute renal failure there is a state of hyperendorphinism.

Kaptein: I have one comment and one question. It is intriguing to link the change in thyroid hormone metabolism to those of the gonadal system. However, I was previously totally unaware that there had been any measurements of TRH which, of course, would have to be from the hypothalamic portal circulation in patients with low T_3, or in hypothyroidism that were convincing. Secondly, the T_4 production rates, of course, have never been shown to be decreased in these patients. However, it is intriguing — I am just not aware of the data. The question I had was regarding the correlations of testosterone and LH. I suspect these patients could have changes in testosterone-binding protein concentrations and therefore I wondered whether you looked at free testosterone and whether that related to the LH levels in a more direct manner than total testosterone?

Kokot: Yes, it is a very important point but we did not estimate free testosterone. Therefore, I cannot correlate these two factors.

Massry: Dr. Kaptein, you know from our data that you were initiating, that we didn't find free testosterone in acute renal failure. Despite a normal total testosterone level and testosterone protein binding, the free testosterone concentration was low. I must say that we did not find a correlation between testosterone and PTH in acute renal failure. The levels are very low and therefore you could not correlate. We postulated that the changes in the gonadal hormones are related to changes in the calcium content of the critical endocrine areas of the testes. Therefore, performing studies in parathyroidectomized and non-parathyroidectomized dogs with acute renal failure and in normal dogs receiving excess PTHh, we found that the hypothalamus, the pituitary gland and the testes have excess calcium whether they have renal failure or not, as long as there is excess PTH. They do not have high calcium, even in the presence of acute renal failure and uremia if there is no excess PTH. In these three critical areas, indeed, the presence of PTH results in accumulation of calcium. Does this mean anything? When you measured testosterone in the animals with intact parathyroid gland there was a precipitous fall in the level of testosterone after induction of uremia. In contrast, induction of uremia in parathyroidectomized animals did not cause significant changes of testosterone concentration. Then we took normal animals with normal renal function and treated them for 3 days with PTH. As a result we found the same alterations as in the uremic animals. Therefore, we concluded that PTH plays a role in the low testosterone levels in acute uremia.

Kokot: Thank you very much for these comments, but may I ask you a question? Why do you not have hypogonadism in primary hyperparathyroidism?

Massry: I have not measured the testosterone there.

Kokot: From a clinical point of view we never observed hypogonadism in primary hyperparathyroidism.

Massry: The duration may be too short, levels of PTH may not be as high. I do not know.

Bommer: Some years ago I would have agreed with Dr. Kaptein that T_4 levels are often in the lower normal or hypothyroid range. But we had observed a case with severe galactorrhea and extremely high prolactin levels. The underlying disease was hypothyroidism. After correction of the hypothyroidism, the prolactin level fell to the normal range. I think this observation supports your hypothesis that TRH may stimulate prolactin in such patients. In the literature the syndrome was described as a so-called overlap syndrome. About 20 cases of hypothyroidism with hyperprolactinemia and galactorrhea are reported. Therefore, it cannot be generally excluded that low T_4 has an effect in hemodialyzed patients.

Kaptein: Just a quick comment, mainly that free T_4 by equilibrium dialysis and T_4 production rates are not low and really the only thing left would be the low T_3 levels. I agree with you that hypothyroidism mainly related to a low free T_4 and low T_4 production rate is definitely associated with hyperprolactinemia, particularly in sever stages, but a low T_3 per se is causing it? I am totally unaware of this possibility. I say it is intriguing, but I am totally unaware of any data. Low T_4, low free T_4, I do not argue. It is there but it is not present in patients with acute renal failure.

Massry: It is a provocative concept.

Gross: I wanted to ask you whether the concept of the central pathogenetic role of prolactin can be supported by treatment or experimental application of bromocriptine?

Kokot: We have no experience with the long-term administration of bromocriptine in patients with acute renal failure. Are you talking about acute or chonic renal failure?

Gross: Chronic – giving bromocriptine.

Kokot: Yes, even if you are giving bromocriptine at a very low dose, there are usually side-effects which are intolerable and which the patient does not accept. From the clinical point of view, the effect on the gonadal function was not very high. I think you can see increased sexual activity after long-term administration of bromocriptine.

Massry: Dr. Gross, I think that the Heidelberg group, of which you are part, reported 3 or 4 years ago in the *Lancet* that bromocriptine somehow improved impotence and I think also caused a rise in testosterone.

Bommer: The impotence improved.

Contr. Nephrol., vol. 50, pp. 139–152 (Karger, Basel 1986)

Management of Uremic Patients with Sexual Difficulties

Jürgen Bommer

Medizinische Universitätsklinik, Abteilung Nephrologie, Heidelberg, FRG

The well-known sexual problems of patients on maintenance hemo-dialysis have been studied extensively. Despite this, therapy of sexual dysfunction in such patients still remains unsatisfactory. The following review tries to summarize the currently available therapeutic approaches.

Problems of Sexual Disturbances in Dialyzed Male Patients

In the study of Levy [1], 65% of the 287 patients examined were totally or partially impotent, total impotence occurring in 23%. Normal potency was found in no more than 31% (table I).

In retrospective analyses, Abram et al. [2], Thurm [3] and our group [4] found that this frequency of intercourse was reported to be 10–13 times less than before uremia (table II). In end-stage renal failure prior to dialysis American patients reported 5 episodes of intercourse per month whereas our German patients (presumably more realistically) reported only occasional sexual intercourse. During maintenance hemodialysis, Abram et al. [2] and Thurm [3] reported 4 episodes of intercourse per month. In our dialyzed patients, 6.6 episodes of intercourse per month were reported by the patients. This was confirmed by a separate interview of the spouse which showed adequate agreement with the patient's estimate. However, on average 25% less episodes of intercourse were reported by the spouses. With respect to therapy, our observation appears of importance that when dialysis patients had undergone somatic complications, e.g. myocardiac infarction, pericarditis, severe infection, sexual activity was drastically reduced only of recover after several months of relative physical well-being.

Table I. Prevalence (%) of impotence, partial or total, in male patients: after Levy [1]

	Hemodialysis (n = 287)	Transplanted (n = 56)
Partial impotence	33	33
Total impotence	23	9
No problem	31	56

Table II. Frequency per month of sexual intercourse of male patients

	Abram et al. [2]	Thurm [3]	Bommer et al. [4]	
			patient	spouse
Before uremia	10.5	8–12	13.5	9.6
Uremia	5.7	–	0	0
Chronic HD	4.0	4	6.6	4

Prevalence of Sexual Disturbances in Dialyzed Female Patients

For obvious reasons, information on dialyzed female patients is less complete. Levy [1] observed diminution of sexual activity with progression of renal failure. 33% of women on maintenance hemodialysis reported no sexual activity and 44% only one episode of intercourse per week or less (table III). While 63% of women reported regular orgasm during sexual intercourse prior to uremia, only 33% of chronically dialyzed women had regular orgasm during intercourse. This figure may not be too accurate since 32% of the patients did not answer the questions in the study of Levy [1].

On a clinical level, menstrual abnormalities are more troublesome. They may contribute to anemia of dialyzed women. 73% of female patients with preterminal renal failure still had menstrual bleeding while the proportion decreased to 66% in women in terminal renal failure and to 39% in women on maintenance hemodialysis [5]. Of those women who had menstruation, only 45% had regular menstruation while 55% suffered from hypermenorrhea, metrorrhagia, oligorrhea, etc. [5–7] (table IV).

Table III. Frequency (%) per week of sexual intercourse of female patients (n = 142): after Levy [1]

	Before uremia	Hemodialysis
Never	9	33
Once or less	41	44
Twice	30	11
Three or more	15	6
No answer	5	7

Table IV. With progressing renal failure, disturbances of menstruation occur more frequently: after Huriet et al. [5]

	n	Regular menstruation, %	Irregular menstruation, %
Before uremia	138	81	19
Uremia creatinine 10 mg/dl	93	68	32
Hemodialysis	53	45	55

Exploration of Male Dialysis Patients with Sexual Disturbances

Undoubtedly, there is an organic basis for sexual dysfunction in uremia [8]. However, sexual function in humans is complex and depends on the interaction of various systems at different levels of organisation. Psychic problems, behaviorial disturbances, derangements of supragonadal and gonadal hormones as well as target organ disturbances may interfere with sexual function. As a consequence, one should carefully consider a number of possible pathogenic factors when dialyzed patients complain about disturbed sexual function.

Table V provides a checklist of items to be assessed when caring for such patients. First of all, when patients complain about diminishing sexual drive, one has to consider that patients when they have reached the stage of hemodialysis are older than in the preuremic stage, so that loss of sexual activity with age must be considered. Alcoholism and liver disease are not uncommon in dialysis patients. 20% of our dialysis patients are diabetics who may suffer from disturbed potency on that account. The role of arterioscle-

Table V. Factors other than uremia which can disturb potency in male patients

Age	Drugs
Liver disease, alcoholism	Neuroleptics
Diabetes mellitus	Chlorpromazine
Vascular diseases	Triflupromazine
Atherosclerosis	Thioridazine
Claudication	Antidepressive medication
Polyneuropathy	Lithium
Pelvic surgery	Tricyclics
	Tranquilizers
	Diazepam
	Hypnotics
	Barbiturates
	Antiepileptics
	L-Dopa
	Cimetidine, ranitidine
	Antihypertensive drugs
	Reserpine
	Clonidine
	Methyldopa
	Guanethidine
	β-Blockers

rosis in the genesis of disturbed potency has certainly been underestimated. The penile pressure necessary for vaginal penetration must be in excess of 75 mm Hg [9]. This is far above the level of pressure which would cause claudication in the lower extremity, if similar vascular stenoses and pressure drop across stenoses were present. Consequently, disturbance of potency may be a more sensitive index of vascular obstruction than claudication.

Polyneuropathy is frequently mentioned as a cause of impotence. Classical symptoms for neuropathic impotence are impaired cremaster reflex and bulbocavernosus reflex. This is not commonly noted in the impotent dialysis patients whom we examined. One should also realize that even minor pelvic surgery, e.g. surgery of hemorrhoids, may damage the pelvic nerves and be a cause of impotence.

One must also consider the possibility of impairment of sexual function by drug treatment. Schaefer and von Herrath [10] noted that in a sample of German dialysis patients many patients received drugs which potentially interfere with sexual function. Examples include neuroleptics (e.g. chloropromazine), antidepressants (e.g. tricyclic agents), tranquilizers (e.g. diazepam), hypnotics (e.g. barbiturates) and, above all, antihypertensive agents.

Diagnostic Procedures in Male Dialysis Patients with Sexual Disturbances

A widely used tool to distinguish psychic and organic causes of impotence is the phallograph, i.e. a strain transducer to document the occurrence of nocturnal penile tumescence [11, 12]. Regular noctural penile erections tend to occur in patients with psychic causes of impotence, but not in those with organic causes [12]. When impotent dialysis patients are examined with the phallograph, virtually all have impaired nocturnal penile tumescence in the experience of Procci et al. [12].

However, to exclude added organic causes, some additional procedures may be useful. Impaired vascular inflow into the corpus cavernosum via the dorsal penile and profound penile arteries can be detected by Doppler techniques with or without sphygmometric measurements [13]. Using these techniques, it could be shown in nonuremic patients that even minor arterial occlusions seriously impair maximal penile pressure. Virag et al. [13] found arteriosclerosis of visceral pelvic arteries in a surprisingly large proportion of apparently healthy males, the proportion increasing with age. Latent pelvic artery stenoses may also explain the frequent occurrence of impotence in hypertensive patients treated by diuretics, since compensatory sympathetic vasoconstriction superimposed upon latent stenoses may critically reduce arterial inflow velocity. Conversely, impaired venous drainage can be detected by phlebography. As screening test to detect vascular perfusion problems it has been proposed to measure the temperature difference between the urethral mucosa and sublingual mucosa [14]. A difference of more than 1.7 °C is indicative of impaired arterial perfusion of the penis.

Uremic patients tend to have low testosterone and high LH levels. However, no tight correlation is found between sexual disturbance and gonadal hormone concentrations. In addition, testicular volume tends to shrink in patients after several years of dialysis. Although there is some information on testicular morphology, the causes of late testicular atrophy have not been well defined. Obviously, severe testicular atrophy will determine the kind of therapeutic intervention planned. A substantial pro-

Table VI. Therapeutic approaches in male dialysis patients with sexual disturbances

---

Psychological methods
Testosterone
Gonadotropins
  hCG
  LH, FSH
1,25-Dihydroxy vitamin D_3
Zinc
Prolactin inhibitors
  Bromocriptine
  Lisuride
Clomiphene citrate

---

portion of dialyzed patients have elevated human prolactin (hPRL) levels [15]. Again, neither sexual dysfunction is predicted by hPRL levels nor therapeutic success with dopaminergic agonists, so that therapeutic decisions should not be based exclusively on such hormone measurements [16].

Therapeutic Approaches in Male Dialysis Patients with Sexual Dysfunction (table VI)

Before any intervention it is recommended to have a separate interview with the patient and his partner. It is imperative to sensibly explain to the partner the patient's problem. In particular, the partner should avoid confronting the patient with demands that cannot be fulfilled. Performance anxiety ('Versagensangst') is a common problem according to our experience. Much is gained when the partner is advised to leave the initiative in sexual activity exclusively to the diseased male partner.

It may even be of help to strictly forbid genital intercourse. Human curiosity will tend to overcome such medical advice (which of course is the reason why such advice is given in the first place). Sometimes it is useful to advise couples to avoid genital intercourse but to try nongenital play thus having a better chance of experiencing orgasm. This will help to give confidence to the diseased male patient and help to overcome performance anxiety.

Hormonal treatment with testosterone is not only of unproven efficacy, but even has definite hazards, e.g. priapism and liver tumors. Testosterone may also be disadvantageous since it tends to increase libido without concomitant increase of potency, thus heightening the patient's dilemma. Human chorionic gonadotropin (Primogonyl®) has been shown to raise testosterone levels. Long-term success with this treatment has not been reported. This may reflect either that long-term parenteral treatment is impractical or that no relation exists between circulating testosterone levels and sexual dysfunction. Some success has been reported with the estradiol analogue clomiphene [17] but again, no long-term experience on efficacy and safety is available.

Both PTH excess [18] and deficiency in $1,25(OH)_2$ vitamin D_3 [19] have been implicated in gonadal dysfunction of uremia. Indeed, treatment with $1,25(OH)_2$ vitamin D_3 has been shown to raise testosterone levels but clearcut effect and benefit on sexual dysfunction has not yet been demonstrated. Under these circumstances, however, one should correct abnormalities of calcium metabolism, if present, in patients with sexual dysfunction.

Conflicting results have been reported with respect to zinc therapy. Antoniou et al. [20] and Mahajan et al. [21] found improvement of sexual function in hemodialyzed patients when zinc preparations were administered per os. Administration of zinc was based on the consideration that zinc levels in the gonads are high and that zinc deficiency causes acquired hypogonadism and testicular atrophy. However, in a controlled double-blind study, no beneficial effect of zinc therapy was found on libido, nocturnal penile tumescence, serum testosterone, LH or FSH levels [22]. Such negative results were confirmed by other investigators [23–25]. It is unclear to what extent such conflicting results can be explained by differences in the duration of the study, circadian changes of steroids or differences in zinc dose or route of application.

One therapeutic approach with proven efficacy is the administration of dopaminergic agonists, e.g. bromocriptine (2.5–5.0 mg/day) or its more recent congener Lisurid (lisuride hydrogen maleate, 0.05–0.2 mg/day). The use of these agents is based on the consideration that in nonuremic patients elevated prolactin levels are frequently associated with impotence [26], such impotence responding to suppression of prolactin secretion by dopaminergic agonists [26, 27]. It is important that hPRL levels prior to and during treatment do not predict success of therapy [16].

In 1979 we reported successful administration of bromocriptine in 7 patients [28]. This was associated with improved potency and sexual activity in 6 patients; serum testosterone levels increased in 4 patients, although the latter was not statistically significant. Similar results were reported by other authors in a double-blind study [16]. Unfortunately, bromocrip-tine has potentially severe side-effects. The most notable are ones nausea and lowering of blood pressure. The latter may lead to fistula clotting according to our experience [unpubl.]. Such side-effects are considerably less frequent and less pronounced with Lisurid. In particular, the first dose effect, i.e. an unpredictable blood pressure fall upon initiation of therapy, is uncommon.

Treatment of Impaired Fertility of Dialyzed Male Patients

Some information is avalaible on treatment of disturbed libido and potency. In contrast, very little information is avalaible with respect to treatment of impaired fertility.

In hemodialyzed patients, sperm counts and sperm motility are markedly diminished [17, 29] and there is little or no correlation with circulating FSH and testosterone levels.

Any effort to treat infertility must address the nonmedical, but very important issue whether it is desirable for the patient to sire children when life expectancy is unavoidably uncertain. We tend to advise our patients to postpone family planning until a successful renal transplant has been in place for at least 2 years or until the patient has been stable on dialysis without major complications for prolonged periods of time.

If after a sympathetic discussion of these aspects with the patient he insists, the following measures can be offered:

Cohabitation should be restricted to the time of maximal probability of conception, since reduction of frequency of cohabitation will increase the chance to obtain many viable sperms. It is unwise, as done by others, to advise frequent intercourse at the expected time of ovulation because this exhausts spermatogenesis and may exhaust sperm counts at the critical moment. Exposing testes to elevated temperature, e.g. tight-fitting pants or Japanese baths, should be discouraged. In patients with erectile impotence, electrostimulation to promote ejaculation or recovery of sperm with subsequent artificial insemination has been recommended. It should be left to the patient and his treating physician to decide whether one should go thus far. There is little evidence that hormonal interventions improve spermatogenesis with the exception of the report of Lim and Fang [17] who reported a small increase of sperm counts (but no increase of motility!) after treatment with clomiphene citrate.

Treatment of Sexual Disturbances in Dialyzed Females

Menstrual disturbances in females may be a problem of considerable practical clinical importance because menstrual bleeding may contribute to anemia. Most dialyzed females have anovulatory cycles with impaired or absent progesterone burst and inadequate luteal transformation of the endometrium, giving rise to prolonged and irregular bleeding [30]. In some cases, early abortion may underlie abnormal gynecological bleeding. That is apparently quite common in dialyzed female patients since Huriet et al. [5] found increased rate of abortion and impaired nidation in advanced renal failure.

If a female patient desires to have menstrual bleeding, adequate endometrial transformation can be achieved by administration of luteal hormones during the second half of the cycle (14th–25th day), e.g. 4 mg medroxyprogesterone/day.

Alternatively, if bleeding has to be avoided for personal reasons or for prevention of anemia, endometrial atrophy can be induced by continuous high dose progesterone, e.g. 25–100 mg medroxyprogesterone/day per os, or by intermittent intramuscular injection of depot progesterone, e.g. 150 mg medroxyprogesterone acetate every 2 months. Such continuous high dose progesterone therapy is not free of side-effects, particularly nausea.

For female dialysis patients who want to conceive, the same general considerations apply as discussed above for males. Because of the important role of the mother in the development of a child, very careful consideration should be given to the potential life expectancy of the prospective mother. It is our opinion that pregnancy should be avoided in the interest of the prospective child. In theory, conception can be facilitated by administration of bromocriptine which has been shown to transform anovulatory cycles into ovulatory cycles [31].

References

1   Levy, N.B.: Sexual adjustment of maintenance hemodialysis and renal transplantation. National survey by questionnaire, preliminary report. Trans. Am. Soc. artif. internal Organs *19:* 138–143 (1973).
2   Abram, H.S.; Hester, L.R.; Sheridan, W.F.; Epstein, G.M.: Sexual functioning in patients with chronic renal failure. J. nerv. ment. Dis. *160:* 220–226 (1975).
3   Thurm, J.: Sexual potency of patients on chronic hemodialysis. Urology *5:* 60–62 (1975).

4   Bommer, J.; Tschöpe, W.; Ritz, E.; Andrassy, K.: Sexual behaviour of hemodialysis patients. Clin. Nephrol. *6:* 315 (1976).

5   Huriet, C.; Mire, F.; Kessler, M.; Le Fall, E.; Gauthier, P.; Mur, J.-M.; Landes, P.; Grignon, G.; Richon, J.: Profil clinique et cytologique de la femme en hémodialyse. J. Urol. Néphrol. *4−5:* 369−375 (1974).

6   Morley, J.E.; Distiller, L.A.; Epstein, S.; Katz, M.; Gold, C.; Sagel, J.; Kaye, G.; Pokroy, M.; Kalk, J.: Menstrual disturbances in chronic renal failure. Hormone metabol. Res. *11:* 68−72 (1979).

7   Rice, G.G.: Hypermenorrhea in the young hemodialysis patient. Am. J. Obstet. Gynec. *116:* 539−543 (1978).

8   Holdsworth, S.R.; De Kretser, D.M.; Atkins, R.C.: A comparison of hemodialyse and transplantation in reversing the uremic disturbance of male reproductive function. Clin. Nephrol. *10:* 146−150 (1978).

9   Virag, R.: Arterial and venous hemodynamics in male impotence; in Bennet, The management of male impotence, pp. 108−125 (Williams & Wilkins, Baltimore 1982).

10 Schaefer, K.; Herrath, D. von: Personal commun.

11 Leliefeld, H.H.J.; Mühr, S.: Erectile impotence − clinical experience with the phalloplethysmography. Urol. int. *37:* 258−266 (1982).

12 Procci, W.R.; Goldstein, D.A.; Adelstein, J.; Massry, S.G.: Sexual dysfunction in the male patient with uremia. A reappraisal. Kidney int. *19:* 317−323 (1981).

13 Virag, R.; Bouilly, P.; Frydman, D.: Is impotence an arterial disorder? A study of arterial risk factors in 440 impotent men. Lancet *i:* 181−184 (1985).

14 Becker, H.C.; Weidner, W.; Krause, W.; Rothauge, C.F.: Zur Diagnostik der Impotentia erectionis aus somatischer Sicht. Diagnostik *15:* 1194−1203 (1982).

15 Gomez, F.; De la Cueva, R.; Wauters, J.-P.; Lemarchand-Béraud, T.: Endocrine abnormalities in patients undergoing long-term hemodialysis. The role of prolactin. Am. J. Med. *68:* 522−530 (1980).

16 Muir, J.W.; Besser, G.M.; Edwards, C.R.W.; Rees, L.H.; Cattell, W.R.; Ackrill, P.; Baker, L.R.I.: Bromocriptine improves reduced libido and potency in men receiving maintenance hemodialysis. Clin. Nephrol *20:* 308−314 (1983).

17 Lim, V.S.; Fang, V.S.: Restoration of plasma testosterone levels in uremic men with clomiphene citrate. J. clin. Endocr. Metab. *43:* 1370−1377 (1977).

18 Massry, S.G.; Goldstein, D.A.; Procci, W.R.; Kletzky, O.A.: Impotence in patients with uremia. A possible role of parathyroid hormone. Nephron *19:* 305−310 (1977).

19 Kreusser, W.; Spiegelberg, U.; Ritz, E.: Wirkung von $1,25(OH)_2D_3$ auf den hypergonadotropen Hypogonadismus bei Niereninsuffizienz. Nieren- Hochdruckkrankh. *8:* 237 (1979).

20 Antoniou, L.D.; Sudhaker, T.; Shalhoub, R.J.; Smith, R.C.: Reversal of uremic impotence by zinc. Lancet *ii:* 895−898 (1977).

21 Mahajan, S.K.; Prasad, A.S.; Briggs, W.A.; McDonald, F.D.: Effect of zinc therapy on sexual dysfunction in hemodialysed patients. Trans. Am. Soc. artif. internal Organs *26:* 139−141 (1980).

22 Brook, A.C.; Johnston, D.G.; Ward, M.K.; Watson, M.J.; Cook, D.B.; Kerr, D.N.S.: Absence of a therapeutic effect of zinc in the sexual dysfunction of hemodialysis patients. Lancet *ii:* 618−619 (1980).

23 Schäfer, M.; Mies, R.; Vlaho, M.: Zinc substitution for male dialysis patients: positive

effect on preexisting hypogonadism? Contr. Nephrol., vol. 38, pp. 116–118 (Karger, Basel 1984).

24 Sprenger, K.B.G.; Schmitz, J.; Hetzel, D.; Bundschu, D.; Franz, H.E.: Zinc and sexual dysfunction. Contr. Nephrol., vol. 38, pp. 119–125 (Karger, Basel 1984).

25 Zetin, M.; Stone, R.A.: Effects of zinc in chronic hemodialysis. Clin. Nephrol. *13:* 20–25 (1980).

26 Franks, S.; Jacobs, H.S.; Martin, N.; Nabarro, J.D.N.: Hyperprolactinemia and impotence. Clin. Endocr. *8:* 277 (1978).

27 Thorner, M.O.; McNeilly, A.S.; Hagan, C.; Besser, G.M.: Long-term treatment of galactorrhea and hypogonadism with bromocriptine. Br. med. J. *ii:* 419 (1974).

28 Bommer, J.; Ritz, E.; Pozo, E. de; Bommer, G.: Improved sexual function in male hemodialysis patients on bromocriptine. Lancet *ii:* 496–497 (1979).

29 Gupta, D.; Bundschu, H.D.: Testosterone and its binding in the plasma of male subjects with chronic renal failure. Clinica chim. Acta *36:* 479–484 (1972).

30 Lim, V.S.; Henriquez, C.; Sievertsen, G.; Frohman, L.A.: Ovarian function in chronic renal failure. Evidence suggesting hypothalamic anovulation. Ann. intern. Med. *93:* 21–27 (1980).

31 Wass, V.J.; Wass, J.A.H.; Rees, L.; Edwards, C.R.W.; Ogg, C.S.: Sex hormone changes underlying menstrual disturbances on hemodialysis. Proc. Eur. Dial. Transplant Ass. *15:* 178–186 (1979).

Prof. Dr. J. Bommer, Medizinische Universitätsklinik, Abteilung Nephrologie, Bergheimer Strasse 58, D–6900 Heidelberg (FRG)

Discussion

Gurland: Thank, you, Dr. Bommer, for a very clear and detailed overview. Are there any differentiations in your experience or in the literature in how far different modes of dialysis treatment differ in sexual outcome of the patients? For instance, may be CAPD has better success in treatment?

Bommer: There are only a few reports on CAPD patients. I think other additional problems have to be considered in CAPD patients like the abdominal transcutaneous catheter. The few reports on impotence in CAPD patients show similar results as were found in hemodialyzed patients.

Brodde: What percent of impotence is due to psychological reasons and what is the percent really due to uremia?

Bommer: It is difficult to differentiate between these two causes of impotence; perhaps Dr. Massry can give us some more information. He has measured nocturnal penile tumescence in his patients, we have not done this systematically, but many of our patients report of erection in the morning. I think this symptom suggests that psychological factors are very important, e.g. stress has to be considered. Each dialysis session is a stress situation for the patient. Some patients have social problems with their family or their friends. Some patients are concerned about their job. Often, it is difficult to realize the importance of psychological and organic factors. If I try to estimate the relevance of both, I think the psychological problems dominate in about 40% of the cases.

Brodde: As you know, a lot of dialysis patients, independent of age, are unemployed. Sometimes they get the chance to get a job. Is there any improvement after changing that situation?

Bommer: We have not analyzed our data for this question systematically, but I have the impression that patients who have regained a job have less sexual problems. Only a few patients, perhaps 5, were unemployed and got a job again. I think the number of patients is too small to answer this question.

Massry: I really find difficulty in defining what psychological factors are; it is just a vague term that we use and we all think that every patient with uremia has psychological troubles. I think psychiatrists have accepted the fact that depression may cause impotence in a nonuremic patient or in an uremic patient. Therefore, in our studies of a large population of nondialyzed and dialyzed patients, we evaluated, through a psychiatrist, depression in the patient, using three different batteries of tests. I must say that we found neither correlation between the presence or absence of depression and impotence as manifested by the information the patient delivered about his frequency of intercourse, nor any correlation with the magnitude of penile nocturnal tumescence. So we concluded from that study that depression probably does not play a major part in the impotence of the uremic patient. Now if we take penile nocturnal tumescence as a criterion to differentiate between 'psychological impotence' and 'organic impotence', we found that 50% of uremic dialyzed and nondialyzed patients have reduced penile nocturnal tumescence, which corresponds to the percentage of decreased sexual intercourse as delivered by the patient. We further found, and that is very interesting, that there is a direct and very significant correlation between the severity of autonomic neuropathy and the frequency of sexual intercourse in patients who have active sexual partners. Frequency of intercourse is a very difficult thing to interpret. If people do not have available active sexual partners, of course the frequency of intercourse may vary. In patients who do have available active partners for sexual intercourse, there was a strict and direct and significant correlation between autonomic neuropathy and intercourse. So we believe there is more to the organic nature of impotence in the uremic patient than psychological disturbance.

Now to the question. You did not mention the issue of penile implants. This is now very fashionable in the United States and urologists are urging many patients to have penile implants with water pumps which can allow them to have an erection at any time they want and maintain it as long as they want, as long as they maintain the pressure in the pump.

Bommer: In response to your question, we have no experience of penile implants. In the literature there are some reports of vascular reconstruction operation and they have succeeded. I have no experience and this work was not done in uremic patients. Therefore, I cannot answer your question on the possible results in dialysis patients. Our patients did not like this method up to now.

Ritz: There is one point where I strongly disagree. That is with the reliability of psychiatric test batteries to predict anything about depression. I am co-author in a study where our patients were analyzed by our psychosomatic colleagues. Given the same data, I was able to completely reverse their interpretation, so I am not very convinced about the scientific solidity of such data. Coming back to your comment about autonomic polyneuropathy. Dr. Bommer made the point that there is no correlation in the patients in our care. We do not apply many sophisticated tests, but I would take your point that there may be some component. Specifically, we have to consider when talking about erection and potency that there is much more to nervous function than just the catecholaminergic and cholinergic systems. Probably the most important system in erection is the vascular intestinal peptide. Vascular peptidergic nerves play a key role in tumescence and this has never been properly analyzed. I think we still need a reliable test to evaluate these noncatecholaminergic functions.

Gross: It has been shown now that the occurrence of nocturnal penile tumescence is still compatible with a correctible cause of impotence. That does not rule it out. This test is not reliable. Secondly, in an impotent male on antihypertensives, do you try to switch the antihypertensives, and to which?

Bommer: To the first point there are reports that nocturnal penile tumescence is a very reliable measurement. To answer the second question yes, for example I have changed some weeks ago the hypertensive drugs in a patient who had problems on clonidine therapy. This patient got a vasodilator.

Massry: This is in answer to Dr. Gross. I must feel that you are a little bit harsh on that test. I do not think that if it is abnormal, it means there is an irreversible cause of impotence. The only thing they said was that there is an organic cause, and an organic cause is often correctible. I am not sure that, because of what you said, it is an unreliable test. I must tell you also that the impairment in penile nocturnal tumescence correlated excellently with the information the patient delivered about sexual intercourse. As you know, penile nocturnal tumescence occurs at night when we are asleep. Sexual intercourse occurs when we are awake. So if one correlates what happens when we sleep, with what happens when we are awake, and they fit together, it seems that what we have in our sleep is probably a good reflection of what we do when we are awake.

Gross: I wish to avoid a misunderstanding on this. It has been published, I think, in the *Journal of Urology* last year, that men who reported impotence and did show positive nocturnal penile tumescence were, in 50% of the cases, still found to have a correctible cause that abolished their impotence after treatment. So I am only trying to point out that the finding of penile nocturnal tumescence does not necessarily, if this report is to be believed, rule out impotence.

Ritz: Just a brief comment to the practically important question of which antihypertensive medication causes least interference. It has been claimed that calcium antagonists are much better in this respect and, since I see most patients in our Outpatient Department, I have used it systematically whenever there were complaints. I am singularly unimpressed by the efficacy of this. If I may advance a personal hunch and bias, I am very much impressed by the paper in *The Lancet* that Dr. Bommer quoted about vascular lesions which are sufficient to reduce vaginal penetration pressure and I think this is a direction we have to take to exclude such minor lesions and to measure directly the pressure of the penis during erection. I am not convinced that any hypertensive drug is superior to the other.

Heidland: We should know that even calcium antagonists theoretically may impair the calcium-dependent testosterone secretion. However, I am not aware of any clinical investigation on this problem as yet. Another question concerns the place of physical activity in the treatment of impotence. Do you know whether in the study of Goldberg performed in RDT patients the influence of exercise on sexual activity was investigated? There are some hints in the current literature that in healthy individuals heavy physical exercise will reduce sexual activity.

Bommer: I have not got experience in dialysis patients. I have not found a report in the literature. If the patients have to work a lot and have less sleep, only a few hours per night, potency can be decreased and if they have physical stress, then the potency can be diminishes, too.

Brodde: I would not agree with you that exercise may not have good results. Perhaps you should know that there are a lot of studies that long-term exercise or continued exercise can improve sympathetic activity, especially responsiveness of adrenergic receptors. If the sympathetic nervous system is mainly involved in potency, one could speculate that indeed a continued once-a-week or twice-a-week exercise could improve it. I would not agree that you strictly say that there should be no chance, as you did.

Bommer: I have no data, I don't like to speculate.

Contr. Nephrol., vol. 50, pp. 153–166 (Karger, Basel 1986)

Neutral Steroid Metabolites in Body Fluids of Patients with Uremia and after Renal Transplantation[1]

H.V. Henning, Helga Ludwig-Köhn

Division of Nephrology, Department of Internal Medicine,
University of Göttingen, Göttingen, FRG

In patients with chronic renal insufficiency and under regular hemo-dialysis or hemofiltration treatment numerous endocrine dysfunctions and hormonal abnormalities may develop [16]. Little is known, however, about steroid hormones and their metabolism in the various states of renal failure. It has long been recognized that in other diseases which affect steroid biosynthesis and/or metabolism the steroid excretion patterns are considerably altered [5, 23, 29, 32]. In 1978, we showed for the first time that steroid profiles in the urines of patients with chronic renal insufficiency diverge significantly from those of healthy subjects [24]. In the following years investigations by gas chromatography-mass spectrometry (GC-MS) on neutral steroid metabolites in body fluids from uremic patients have been continued in order to extend our previous findings and to possibly gain a better understanding of the wide spectrum of hormonal metabolic disorders in uremia.

Patients and Methods

Sample Collection
Patient groups, biological fluids, sex, age, serum creatinine values and duration of hemo-dialysis (HD), hemofiltration (HF), continuous ambulatory peritoneal dialysis (CAPD) or time after transplantation, respectively, are summarized in table I. 24-Hour urine was collected from 20 healthy individuals (group 1: 6 males, 4 females, mean age 28 and 30 years; group 2: 6 males, 4 females, mean age 55.2 and 58.2 years), from 26 nondialyzed patients with chronic renal failure and with plasma creatinine ranging from 1.3 to 10.0 mg/dl (group 3: 14 males, 12 females, mean age 54.1 and 59.0 years), from 13 uremic patients on regular

[1] Supported by the Deutsche Forschungsgemeinschaft.

Table I. Patient groups, biological fluids, sex, number, age, serum creatinine (mg/dl), and duration of HD, HF and CAPD or time after transplantation

Group	Biological fluid	Sex	n	Age years		Serum creatinine mg/dl		Duration of HD, HF or CAPD or time after transplantation, months	
				mean	range	mean	range	mean	range
1 Healthy	urine	m	6	28.0	23–32	1.0	0.9–1.2	–	–
Controls (n = 10)		f	4	30.0	25–33	0.9	0.8–1.1	–	–
2 Healthy	urine	m	6	55.2	50–64	1.1	0.9–1.2	–	–
Controls (n = 10)		f	4	58.2	53–62	1.0	1.0–1.2	–	–
3 Nondialyzed	urine	m	14	54.1	26–74	4.8	1.5–9.5	–	–
uremics (n = 26)		f	12	59.0	41–75	6.1	1.3–10.0	–	–
4 Uremics on RHDT	urine	m	9	49.2	30–72	12.3	10.8–14.7	29.5	12–48
or RHFT (n = 13)		f	4	54.5	29–72	12.0	9.9–15.5	49.3	23–74
5 Allograft re-	urine	m	12	42.4	23–66	2.1	0.9–5.4	18.8	2.0–53
cipients (n = 22)		f	10	46.0	19–60	1.8	1.1–4.3	26.4	4.0–69
6 Uremics on	hemo-	m	6	59.0	35–74	12.2	10.8–12.8	35.2	13–60
RHFT (n = 11)	filtrate	f	5	49.6	29–65	11.5	8.2–15.5	45.2	29–75
7 Uremics on	CAPD	m	7	48.6	30–62	11.2	9.1–16.7	15.4	1.0–39
CAPD (n = 10)	dialysate	f	3	34.0	24–62	11.1	9.9–12.4	11.0	9.0–14

hemodialysis (RHDT) or hemofiltration (RHFT) treatment (group 4: 9 males, 4 females, mean age 49.2 and 54.4 years), and from 22 patients after renal transplantation (group 5: 12 males, 10 females, mean age 42.4 and 46.0 years). Hemofiltrates from 11 patients (group 6: 6 males, 5 females, mean age 59.0 and 49.6 years) and CAPD dialysates from 10 patients (group 7: 7 males, 3 females, mean age 48.6 and 34.0 years) were also collected. The first 10 liters of hemodialysate of a male patient aged 43 years were used for qualitative analysis. All samples were worked up immediately or stored at −10 °C.

Identification and Quantification of Steroids
Figure 1 shows the enrichment and analyses of steroids from biological fluids. The enrichment of steroid metabolites *from urine* was performed according to Pfaffenberger and Horning [28]. Three parallel steroid determinations were carried out each with 10 ml of 24-hour urine. In 2 samples 10 µg 3β-hydroxy-5β-androstan-17-one [5] and 10 µg 3β,17α-21-trihydroxy-5β-pregnen-20-one [25] were added as internal standards. Only 3β-hydroxy-5β-androstan-17-one was given to the third to enable over-laps in the gas chromatogram by natural corticoids and 3β,17α,21-trihydroxy-5β-pregnen-20-one to be recognized. For the separate estimation of

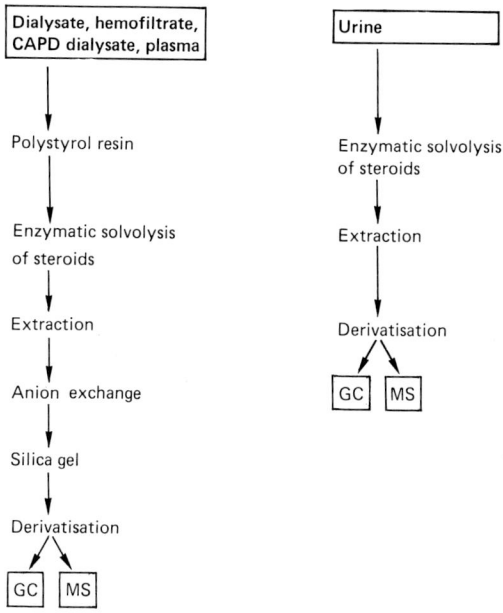

Fig. 1. Enrichment and analysis of steroids from biological fluids.

steroid glucuronides and sulfates 10 ml of 24-hour urine were solvolyzed with either β-glucuro-nidase (20 U, Boehringer, Mannheim, FRG) or arylsulfatase (0.5 U, Boehringer, Mannheim). In order to control the solvolysis of androsterone sulfate which is not completely solvolyzed with arylsulfatase [22] some urine samples were solvolyzed with hydrochloric acid [34]. After trimethylsilylation to give the trimethylsilyl-enol-trimethylsilyl ethers [11] quantification was carried out by capillary gas chromatography [5, 11, 23, 32]. The detection limits of steroid trimethylsilyl ethers were approximately 5–10 ng when splitless injection was used. Standard deviations (SD) of urine steroid investigated in 10 parallel work-up procedures were 5% on average. Correlation measurements with constant standard amounts and increasing amounts of a weighed mixture of 17 steroids that occur naturally in human body fluids revealed a mean correlation factor of 0.98 ± 0.02 SD in 10 determinations with urine as base.

Steroids from *hemofiltrate, hemodialysate and CAPD dialysate* were enriched by adsorption on XAD-2, solvolysis with helicase, extraction with ethylacetate and purified from acids on DEAP-LH-20 and from unpolar substances on silica gel. For quantification 100 μg of each standard were added to 2 liters of hemofiltrate or 2 liters of CAPD dialysate. A detailed description of these work-up procedures has been published elsewhere [25].

Gas Chromatography, Mass Spectrometry

Sources of materials and instruments have been published earlier [25]. Steroids and other compounds which could not be determined by gas chromatography alone were identified by the combination of GC-MS. Gas chromatograms were measured with a Varian 3700 equipped with a fused silica capillary column (Durabond 1.30 m, 0.25 mm inner diameter, 0.25 μm film thickness, ICT Frankfurt, FRG); injection splitless, temperature program, 120–290 °C with 2 °C/min, flow rate 2 ml hydrogen/min, injector and detector temperature 270 °C. For quantification, the peak height of the standards was divided by that of each steroid in the gas chromatograms. This ratio was further multiplied with calculated correction factors to compensate losses during the experimental procedure. The excretion rates were calculated for each individual steroid, then 11-keto- and 11β-hydroxyandrosterone as well as 11-keto- and 11β-hydroxyetiocholanolone were summarized as '11-keto-oxygenated androstanolones' (11-o-Ans), α- and β-cortolone, tetrahydrocortisone, tetrahydrocortisol, and α-tetrahydrocortisol as 'corticoids'.

Results

Figure 2 shows the urinary steroid gas chromatographic profile of a healthy male (group 1), and figure 3 that of a nondialyzed male uremic patient (group 3) at a serum creatinine level of 8.0 mg/dl. Figure 4 shows the profile of the same patient as figure 3 seven months after successful transplantation. The gas chromatographic profiles from hemodialysates and from hemofiltrates were found to be similar to each other and to those in the urine of nondialyzed patients with end-stage renal failure as can be seen in the hemofiltrate profile in figure 5.

Steroid metabolic excretion rates in 24-hour urines of healthy subjects and all patients summarized in table I were measured as well as the elimination rates in 24 h. Two groups of healthy controls were chosen (table I) to represent the extremes of the age range in the patients' groups. The excretion rates of androsterone (A), etiocholanolone (E) and pregnanetriol (PT) declined with age, A showed a significant age-related decrease in healthy males and females (groups 1 and 2). The excretion rates of corticoid metabolites and 11-o-Ans were age-independent. A and E in males of group 3 were in the lower normal range compared to controls of the same age as was the excretion of PT, a metabolite of 17α-hydroxyprogesterone. Excretion

---

Fig. 2. Urinary steroid profile from a healthy male aged 24 years. IS_1 = Internal standard 1 (epietiocholanolone); IS_2 = internal standard 2 (3β,17α,21-trihydroxy-5β-pregnan-20-one); THE = tetrahydrocortisone; CHOL = cholesterol; THF = tetrahydrocortisol; 11-Keto-E = 11-ketoetiocholanolone; 11β–OH-A = 11β-hydroxyandrosterone; 11β-OH-E = 11β-hydroxyetiocholanolone.

Fig. 3. Urinary steroid profile from a uremic male aged 46 years, nondialyzed (serum creatinine = 8.0 mg/dl). For abbreviations, see legend to figure 2.

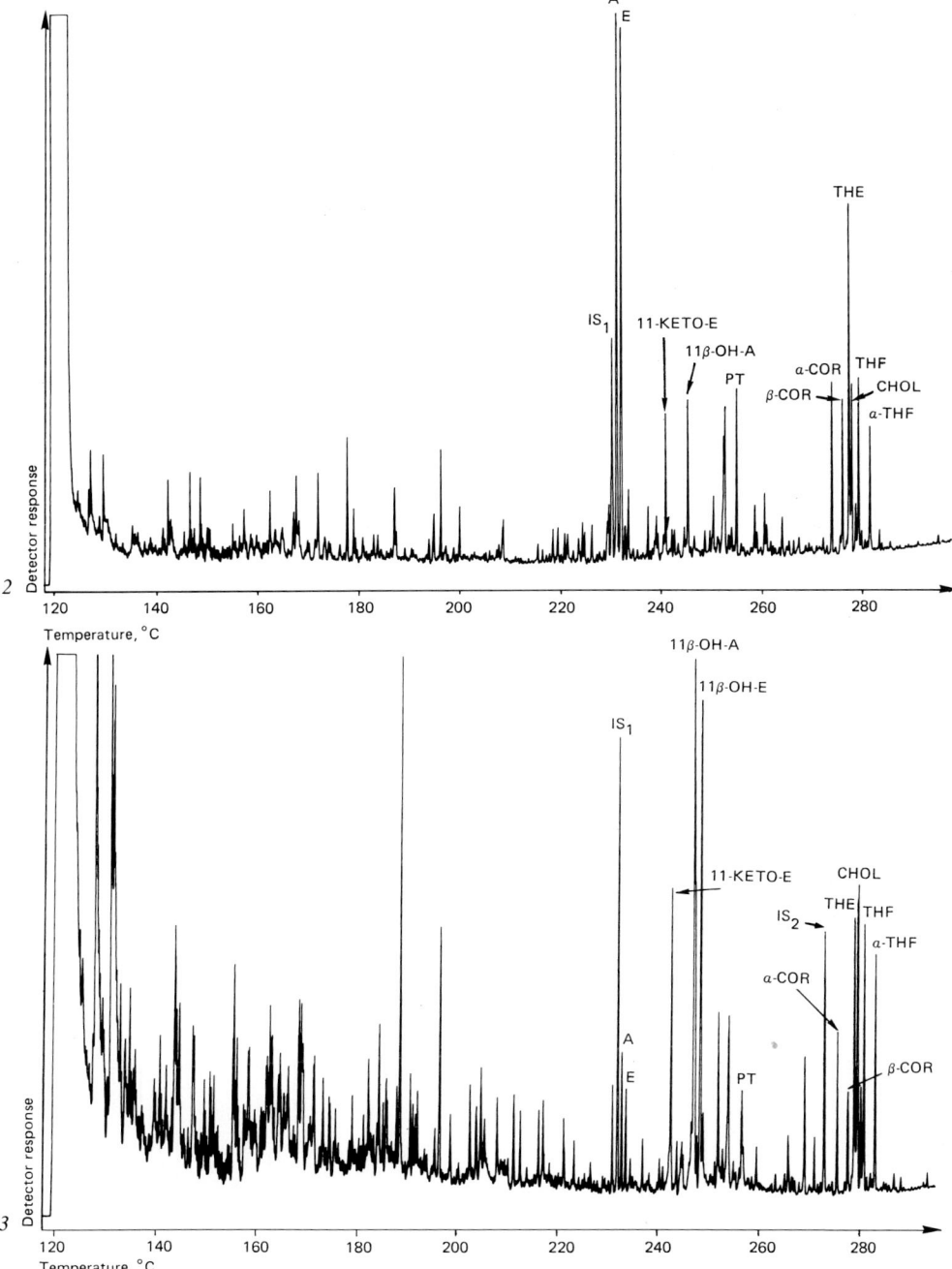

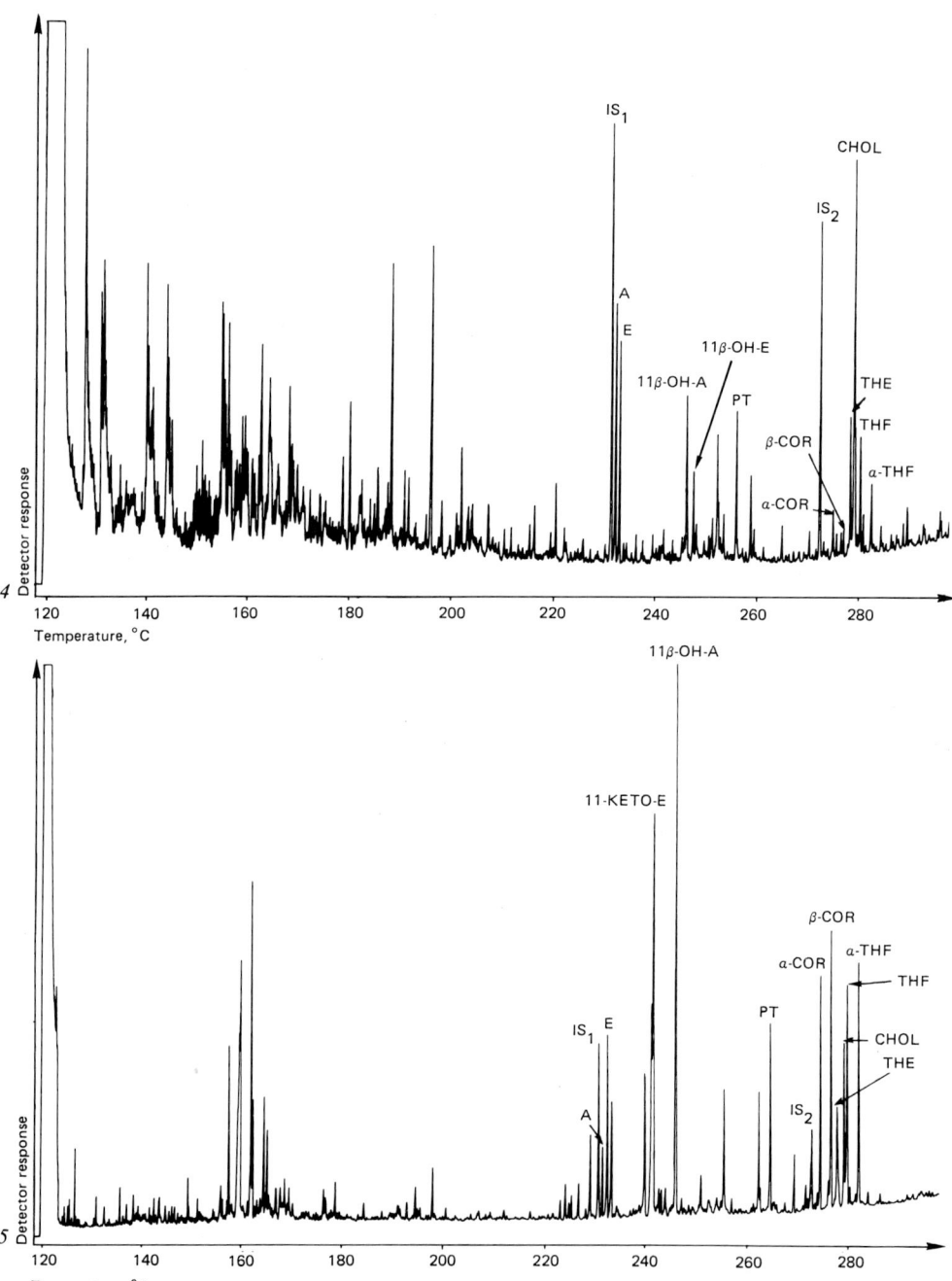

rates of the 11-o-Ans were enhanced up to 200% ($p < 0.05$), and the corti-coid excretion was about 65% of normal ($p < 0.001$). Similar results were obtained for the excretion rates in females: A, E and PT were in the range of the age-matched females in group 2, but the high excretion rate of 11-o-Ans ($p < 0.02$; 250%) and the low rate of corticoid excretion ($p < 0.01$; 65%) were significantly different. The ratios α-cor/β-cor and corticoids/ 11-o-Ans were highly significant in group 3. In all 26 patients of group 3 excretion rates of 11-o-Ans in such an extent have been measured.

Figure 6 shows that correlations exist between the 24-hour excretion of 11-o-Ans and the degree' of renal insufficiency as determined by serum creatinine, and between the ratio corticoids/11-o-Ans and serum creatinine (fig. 7). These correlations proved to be independent of patients' sex and age.

Males on hemofiltration (group 6) eliminated only 30% of A ($p < 0.005$) and 30% of E ($p < 0.001$) compared to the same age group 2, and females 20% of A ($p < 0.005$) and 50% of E (n.s.) compared to group 2. In this group the mean age was higher than that of the females in group 6. The elimination of 11-o-Ans was normal in both sexes while that of the corticoids decreased to 50% ($p < 0.001$).

Separation of glucuronides and sulfates in hemofiltrates and in urines of patients with chronic renal failure revealed that approximately 80% of the steroid metabolites were eliminated as glucuronides and 20% as sulfates. This result is within the normal urinary range.

The elimination rates during 24-hour CAPD (group 7) were in the same range as those under hemofiltration. In comparison to the urinary excretion of healthy individuals (groups 1 and 2), the low elimination rates of steroid metabolites were highly significant in most cases, but in both sexes the elimination of 11-o-Ans was normal. The excretion of steroid metabolites in the residual urines (300–1,500 ml) of dialyzed patients (group 4) during the dialysis- or hemofiltration-free interval decreased further. The urines contained steroid metabolites in amounts which attained half of the hemofiltration elimination rate. After successful renal transplantation (group 5) and treatment with fluocortolone (Ultralan®) the

---

Fig. 4. Urinary steroid profile of the same patient as in figure 3 seven months after successful renal transplantation (serum creatinine = 1.2 mg/dl). For abbreviations, see legend to figure 2.

Fig. 5. Steroid profile of a hemofiltrate, uremic male, aged 62 years. For abbreviations, see legend to figure 2.

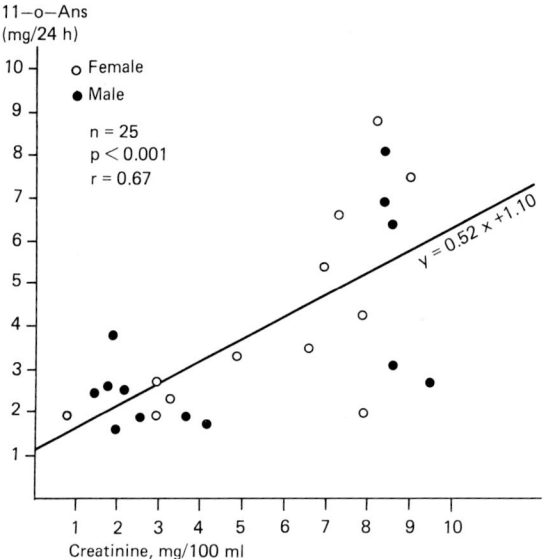

Fig. 6. Correlation of urinary 11-oxygenated androstanolones (11-o-Ans) and serum creatinine in nondialyzed patients with chronic renal failure.

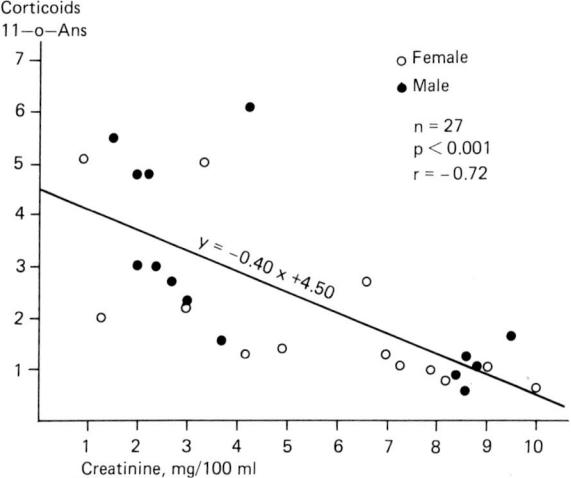

Fig. 7. Correlation of the ratios corticoids/11-o-Ans and serum creatinine in nondialyzed patients with chronic renal failure.

excretion of corticoids was 20% (p < 0.001), the excretion of 11-o-Ans 20−40% (p < 0.001) of normal rates and that of A and E 40−60% compared to group 2 who were on average older than group 5.

Discussion

Avagnina et al. [3] using gas chromatography reported decreased excretion rates of tetrahydrocorticoids, 11-oxygenated androstanolones, A and E in the urine of adult patients with chronic renal failure. We confirmed their results of diminished A, E and corticoid excretion levels (group 3) but in contrast we found elevated 11-o-Ans excretion rates and in nearly all patients a higher E excretion than that of A. Although we cannot reproduce the results of Avagnina et al. [3], we found that in healthy adult males the excretion of A is on average higher than that of E and that in females this ratio is reversed [28, 33]. The elimination rates of A and E in dialysis fluids (groups 6 and 7) were also significantly decreased. The low excretion rates of steroid metabolites in patients with chronic renal failure may depend on diminished production rates of active precursors such as testosterone, estradiol, dehydroepiandrosterone (DHEA) and cortisone or on the reduced glomerular filtration rate of their metabolites. Subnormal to normal plasma levels of DHEA and testosterone [19, 34, 37, 38] and up to high plasma estrogen levels have been found in men [31]. In this study only traces of DHEA could be detected in dialysis fluids. This is surprising in view of the fact that this steroid is water-soluble and dialyzable. After renal transplantation the excretion of A and E does not reach the normal values of healthy individuals due to the intake of fluocortolone at doses of 10−30 mg/day [8]. However, only 4 of the 22 allograft recipients (group 5) had a normal excretory renal function as measured by serum creatinine (0.9−1.2 mg/dl).

Renal tissue contains a β-glucuronidase which is stimulated by testosterone and $3\alpha,17\beta$-dihydroxy-5α-androstane [15, 36]. It is not known whether or not this enzyme, which is responsible for the solvolysis and reactivation of glucuronidated steroids, is inactive in the uremic kidney or cannot be stimulated due to a lack of testosterone. According to our results the metabolism in uremia leads to a weak relative excess of 5β-androstanolones, calculated as the ratio A/E. The degradation of steroids to 5β-androstanolones with loss of their biological activity is not reversible, while the reduction to 5α-products represents a biological transformation [36]. Thus, two mechanisms may possibly be responsible for an in part decreased production of active androgen precursors: deficiency and/or inhibition of the activity of β-glucuronidase and 5α-reductase.

Plasma cortisol levels are usually normal in dialyzed patients but occasionally elevated [1, 4, 26, 27]. The half-life of cortisol is prolonged in chronic renal failure [4], and its metabolism delayed [6, 12, 14]. When the GFR is reduced, the conjugated corticoid metabolites cannot be removed entirely and were found to be accumulated in blood plasma [12, 14]. Englert et al. [14] who found the corticoid metabolite levels decreased to 10% of normal in urines of uremic patients − possibly degraded to artefacts during the experimental procedure − postulated a direct correlation between the reduced GFR and the blood concentration of corticoid metabolites. We could not confirm those low corticoid levels but detected 65% of normal excretion (group 3) which is dependent on sex but after puberty not on age [2]. Deck et al. [12] observed a decrease of 50% of the initial values of plasma corticoid metabolites during hemodialysis that we confirmed indirectly by detecting 50% of normal urinary excretion in the dialysis fluids. These metabolites were also found in hemodialysate by Savard [30].

The healthy kidney transforms cortisone to cortisol [20]. The ratio of the excretion rates of their metabolites represents the biochemical balance of the 11-oxidoreductase system which seems to be disturbed in uremia, shifting to higher tetrahydrocortisol excretion rates [3, 35]. Another disturbance of cortisol metabolism can be observed in the $20\alpha/20\beta$-reductase system: the ratio of α-cor/β-cor (α- and β-cortolone) is significantly decreased in most of our patients. A corresponding enzyme, a 20α-reductase in rat kidney has been described by Hierholzer et al. [21].

An elevated excretion of 11-o-Ans in urine has been reported by Dey and Senciall [13] in hypertensives. However, in our study patients with elevated blood pressure (> 150/90 mm Hg) were only present in group 3 (19 out of 26) but there were no exceptions in their excretion rates. In group 7 (CAPD) only 2 of 10 were hypertensive and in groups 4 and 6 (RHDT and RHFT) hypotension was a major problem in most of the patients. None of our patients suffered from liver disease. Furthermore, as reported earlier we had not found elevated amounts of 11-o-Ans in patients with liver disease [25]. The 11-o-Ans are not artefacts of the experimental procedure or the employment of filters because similar amounts have been detected in dialysis fluids. They may originate from the androgen 11β-hydroxy-4-androsten-3,17-dione [7,9], but probably largely derive from corticoids by loss of the carbon atoms C_{20} and C_{21} [17, 36]. We assume the origin of

11-o-Ans in our study to be corticoids on account of the three following reasons: (1) Corticoid excretion – due to the intake of fluocortolone – and 11-o-Ans excretion decrease simultaneously after transplantation. (2) The excretion rates of 11-o-Ans are independent of age and sex (groups 1 and 2). If they were androgen metabolites, they must decrease with age and be dependent of sex as A and E are. (3) Highly significant correlations exist between the degree of renal insufficiency as determined by serum creatinine and the excretion of 11-o-Ans (group 3) and between serum creatinine and the ratio corticoids/11-o-Ans. Patients with high creatinine levels had low corticoid and high 11-o-Ans excretion rates.

Little is known about the inhibitory effects of 11-o-Ans in biochemical processes. Goldman et al. [18] observed that 11-o-Ans weakly inhibited the production of placental DHEA and an inhibition of the renin-angiotensin system by 11-o-Ans has been reported by Campanile and Goodfriend [10].

The main sites of steroid biochemistry and metabolism are the sexual organs, adrenal cortex and liver [36]. Neutral metabolites in urine provide a typical pattern which can change in diseases affecting steroid biochemistry. Our results were obtained from patients without additional intake of steroid-stimulating or suppressing drugs, except in transplanted patients. The results can be interpreted as being due to disturbances of steroid metabolism in uremia: a relative overproduction of 11-o-Ans originating from cortisol would indicate a shift in activity of 17,20-lyase. Therefore, 11-desoxycortisol, the plasma concentration of which was found to be normal in uremia [1], would scarcely be degraded to A and E [36]. Most of A and E originate from androgen precursors and we found them to be diminished with a relative excess of the 5β-product E in chronic renal failure. 11-o-Ans are metabolic end products and cannot be transformed to A and E [7, 9]. Due to relative excess of tetrahydrocortisol and 20β-cortolone in uremic body fluids the 11-oxidoreductase and the 20α/20β-reductase system may be disrupted.

References

1   Akmal, M.; Manzler, A.D.: Simplified assessment of pituitary-adrenal axis in a stable group of chronic hemodialysis patients. Trans. Am. Soc. artif. internal Organs 23: 703–706 (1977).
2   Aranoff, G.; Roesler, A.: Urinary tetrahydrocortisone and tetrahydrocortisol glucosiduronates in normal newborns, children and adults. Acta endocr. 94: 371–375 (1981).

3   Avagnina, P.; Molino, G.; Cavama, A.; Gariazzo, M.; Gaidano, G.: Il profilo steroidurico gas-cromatografico nell'insufficienza renale cronica. Minerva endocr. 2: 159–164 (1977).

4   Bacon, G.E.; Kenny, F.M.; Murdaugh, H.V.; Richards, C.: Prolonged serum half-life of cortisol in renal failure. Johns Hopkins med. J. 132: 127–131 (1973).

5   Bailey, E.; Fenoughty, M.; Chapman, J.R.: Evaluation of gas-liquid chromatographic method for the determination of urinary steroid using high-resolution open-tubular glass capillary columns. J. Chromat. 96: 33–46 (1974).

6   Boer, P.; Thijssen, J.H.H.: The excretion of metabolites of cortisol and aldosterone in male patients on haemodialysis treatment. Steroids 30: 203–211 (1977).

7   Bradlow, H.L.; Gallagher, T.F.: Metabolism of 11β-hydroxy-Δ^4-androstene-3,17-dione in man. J. biol. Chem. 229: 505–518 (1957).

8   Breuer, H.; Breuer, J.; Luchmann, A.; Schubert, W.: Wirkung von Fluocortolon auf die Ausscheidung von Stickstoff, Steroidhormonen und Elektrolyten im Urin sowie auf die Konzentration verschiedener Serumbestandteile. Fluocortolon-Symp. Berlin 1967, pp. 36–44.

9   Bush, L.E.; Mahesh, V.B.: Metabolism of 11-oxygenated steroids. Biochem. J. 71: 705–717 (1959).

10   Campanile, C.P.; Goodfriend, T.L.: Steroids as potential modulators of angiotensin receptors in bovine adrenal glomerulosa and kidney. Steroids 37: 681–700 (1981).

11   Chambaz, E.M.; Defaye, G.; Madani, C.: Trimethylsilyl etherenol-trimethylsilyl ether. A new derivative for the gas phase study of hormone steroids. Analyt. Chem. 45: 1090–1098 (1973).

12   Deck, K.A.; Siemon, G.; Sieberth, H.G.; Bayer, H. von: Cortisolverlust und Plasma-11-Hydroxy-Corticosteroidprofil während der Haemodialyse. Verh. dt. Ges. inn. Med. 74: 1195–1198 (1968).

13   Dey, A.C.; Senciall, I.R.: Decreased excretion of ring-A reduced steroid metabolites by hypertensive females. Biochem. Med. 12: 213–225 (1975).

14   Englert, E.; Brown, H.; Willardson, D.G.; Wallach, S.; Simons, E.L.: Metabolism of free and conjugated 17-hydroxycorticosteroids in subjects with uremia. J. clin. Endocr. Metab. 18: 36–48 (1958).

15   Fanestil, D.D.; Park, C.S.: Steroid hormones and the kidney. Am. Rev. Physiol. 43: 637–649 (1981).

16   Feldman, H.A.; Singer, I.: Endocrinology and metabolism in uremia and dialysis: a clinical review. Medicine 54: 345–376 (1974).

17   Fukushima, D.K.; Bradlow, H.L.; Hellman, L.; Zumoff, B.; Gallagher, T.F.: Metabolic transformation of hydrocortisone-4-C^{14} in normal men. J. biol. Chem. 235: 2246–2252 (1960).

18   Goldman, A.; Sheth, K.: Inhibitors of human placental C_{19} and C_{21} 3β-hydroxysteroid dehydrogenases. Biochim. biophys. Acta 315: 233–249 (1973).

19   Gupta, D.; Bundschu, H.D.: Testosterone and its binding to the plasma of male subjects with chronic renal failure. Clinica chim. Acta 36: 479–484 (1972).

20   Hellman, L.; Nakada, F.; Zumoff, B.; Fukushima, D.; Bradlow, H.L.; Gallagher, T.F.: Renal capture and oxidation of cortisol in man. J. clin. Endocr. Metab. 33: 52–62 (1971).

21   Hierholzer, K.; Lichtenstein, I.; Siebe, H.; Tsiakiras, D.; Witt, I.: Renal metabolism of corticosteroid hormones. Klin. Wschr. 60: 1127–1135 (1982).

22   Leon, Y.A.; Bulbrook, R.D.; Corner, E.D.S.: Steroid sulfatase, arylsulfatase and
     β-glucuronidase in the mollusca. Biochem. J. *75:* 612–617 (1960).
23   Leunissen, W.J.J.; Thijssen, J.H.H.: Quantitative analysis of steroid profiles from urine
     by capillary gas chromatography. J. Chromat. *146:* 365–380 (1978).
24   Ludwig, H.; Spiteller, G.; Matthaei, D.; Scheler, F.: Profile bei chronischen Erkrankun-
     gen. I. Steroidprofiluntersuchungen bei Uraemie. J. Chromat. *146:* 381–391 (1978).
25   Ludwig-Köhn, H.; Henning, H.V.; Sziedat, A.; Matthaei, D.; Spiteller, G.; Reiner, J.;
     Egger, H.-J.: The identification of urinary bile alcohols by gas chromatography-mass
     spectrometry in patients with liver disease and in healthy individuals. Eur. J. clin. Invest.
     13: 91–98 (1983).
26   McDonald, W.J.; Golper, T.A.; Mass, R.D.; Kendall, J.W.; Porter, G.A.; Girard,
     D.E.; Fischer, M.D.: Adrenocorticotropin-cortisol axis abnormalities in hemodialysis
     patients. J. clin. Endocr. Metab. *48:* 92–95 (1979).
27   Mrinák, J.: Glucocorticoid excretion in long-term dialyzed patients. Acta univ. carol.
     (Med.), Praha *29:* 3–61 (1983).
28   Pfaffenberger, C.D.; Horning, E.C.: Determination of human urinary steroid metabo-
     lites using glass open tubular capillary columns. J. Chromat. *112:* 581–594 (1975).
29   Pike, A.W.; Moynihan, C.; Kibler, S.; Marsh, P.G.; Fennessey, P.V.: Semi-microquan-
     titative analysis of complex urinary steroid mixtures in healthy and diseased states. J.
     Chromat. *306:* 39–50 (1984).
30   Savard, K.: Studies on the steroids in human peripheral blood. Ciba Found. Coll. En-
     docr. *11:* 252–262 (1957).
31   Sawin, C.T.; Longcope, C.; Schmitt, G.W.; Ryan, R.J.: Blood levels of gonadotropins
     and gonadal hormones in gynecomastia associated with chronic hemodialysis. J. clin. En-
     docr. Metab. *36:* 388–390 (1973).
32   Shackleton, C.H.L.; Honour, J.W.: Simultaneous estimation of urinary steroids by
     semi-automated gas chromatography. Investigation of neonatal infants and children with
     abnormal steroid synthesis. Clinica chim. Acta *69:* 267–283 (1976).
33   Sitzmann, F.C.: Normalwerte, pp. 110–117 (Hans Marseille, München 1967).
34   Szücs, J.; Bodrogi, L.: Androsterone sulphate, dehydroepiandrosterone sulphate and
     free dehydroepiandrosterone in uremic patients. Hormone metabol. Res. *12:* 417–418
     (1980).
35   Vanluchene, E.; Vanderkerckhove, D.; Thiery, M.; Van Holder, R.: Changes in cortisol
     metabolism in various physiological and pathological situations. Annls Endocr. *42:*
     284–285 (1981).
36   Traeger, L.: Steroidhormone (Springer, Berlin 1977).
37   Winer, R.L.; Rajudin, M.M.; Skowsky, W.R.; Parker, L.N.: Preservation of normal ad-
     renal androgen secretion in end stage renal disease. Metabolism *31:* 269–273 (1982).
38   Zumoff, B.; Walter, L.; Rosenfeld, R.S.; Strain, J.J.; Degen, K.; Strain, G.W.; Levin,
     J.; Fukushima, D.: Subnormal plasma adrenal androgen levels in men with uremia. J.
     clin. Endoc. Metab. *51:* 801–805 (1980).

Prof. Dr. H.V. Henning, Medizinische Universitätsklinik, Abteilung Nephrologie,
Robert-Koch-Strasse 40, D–3400 Göttingen (FRG)

Discussion

Quellhorst: How could you discriminate between the effects of renal insufficiency and hormonal degradation when you were determining the metabolites in the urine in patients with renal insufficiency?

Henning: What we found is not a hormonal degradation but elevated levels of oxygenated androstanes and the same effect in hemofiltrates, hemodialysates and CAPD dialysates. So we assume that there is an overproduction, and, furthermore, the 11-o-xygenated androstanes are also elevated in serum.

Quellhorst: Do you not believe that there is an interference of the renal excretion in renal insufficiency?

Henning: Yes, we have shown that there is a dependence on serum creatinine. This is a problem which we cannot explain today.

Heidland: This is really an important issue, whether renal metabolism may influence cortisol excretion. I would like to ask you whether it is possible to determine the cortisone metabolites in saliva? There is growing interest in hormonal determinations in saliva, and the results obtained are at least for some hormones representative for the situation in blood. Probably this would be a chance for you to investigate patients prior to dialysis treatment, that means without the use of hemofiltration.

Henning: I do not know any results in the literature from measurements of cortisone in saliva, but I think it would be possible. You can measure it in serum which is somewhat difficult because of high protein contents and other substances, but it should be possible in saliva as well, I think.

Henning: There is another problem which in my opinion is of great importance. That is the question of alterations of serum-protein binding in uremia. We have prepared a so-called 'artificial filtrate' by blood conserves of healthy persons and blood from patients with uremia who underwent plasmapheresis and we found that all metabolites we can measure in serum are also present in the artificial filtrate from the blood conserve of healthy persons but only up to 50% in the artificial filtrate of uremic patients. We have discussed this problem with many outstanding steroid chemists, but they could not help us regarding the problems of altered protein binding of steroids in uremia.

Heidland: Prof. Hierholzer in Berlin, has given convincing evidence that corticosterone and its metabolites are degraded in the rat kidney. Probably in renal insufficiency we have elevated levels, due to impaired renal metabolism. As a clinical consequence, the elevated concentration may be a contributing factor for protein catabolism in the uremic state.

Henning: There is another problem. The steroid metabolism in the kidney of the rat differs significantly from that in humans.

Eigler: With respect to patients with chronic renal insufficiency: Do you know the effect of neutral steroids if they are highly and chronically elevated in these patients for a long time? It could be possible, I think, that they eventually might show some clinically very important effects.

Henning: Until now I do not know. There are findings in the literature that some of these neutral steroids are elevated in patients with hypertension, but we have only some patients with hypertension, most of them are normotensive or hypotensive.

Contr. Nephrol., vol. 50, pp. 167–174 (Karger, Basel 1986)

Renin Substrate (Angiotensinogen) as a Possible Erythropoietin Precursor

F. Fyhrquist, K. Rosenlöf, C. Grönhagen-Riska, V. Räsänen[1]

IVth Department of Medicine, University of Helsinki, and
Minerva Institute for Medical Research, Helsinki, Finland

Erythropoietin (EP) is the main humoral regulator of erythropoiesis. However, both the biogenesis and the mechanism of action of EP are incompletely understood. EP is thought to be produced by kidney [1] or liver [2] of adult, and liver of fetal and neonatal mammals. EP-responsive cells are present in the bone marrow, fetal liver, and adult spleen. EP causes proliferation and maturation of erythroid progenitor cells. According to one hypothesis, kidney is capable of de novo biosynthesis of EP [1]. Another hypothesis maintains that an EP precursor from liver is acted upon by a kidney factor to yield active EP [2].

Gould et al. [3] have suggested an association between the renin-angiotensin system and the production of EP. They have reported data implying a possible role for both renin, angiotensin II, and renin substrate (RS). However, those reports have not allowed final conclusions about the relation between RS and EP to be drawn. We have recently reported immunological similarities between purified human plasma RS and EP, and an erythropoietin-like activity of RS on cultured human erythroleukemic cells of the K562 line [4]. Moreover, we have detected cells in hemangioblastoma tumors which stain clearly positively with antisera against both RS and EP [5]. In 8 renal tumors from patients with high or normal hemoglobin values, neither RS nor EP could be demonstrated with antisera against RS or EP. We here report further investigations exploring the immunological similarities between RS and EP.

[1] Grants were received from the Sigrid Juselius Foundation, and the Nordisk Insulinfond. F.F. received a grant from the von Schoultz Foundation to the memory of Robert Tigerstedt, Finska Läkaresällskapet, K.R. received a grant from the Finsk-Norsk Medicinsk Stiftelse.

Table I. Purification steps of renin substrate (angiotensinogen)

1 NA$_2$EDTA plasma
2 (NH$_4$)$_2$SO$_4$ precipitation
3 DEAE-Sepharose 6 B ion exchange chromatography
4 Blue sepharose chromatography
5 Ultrogel ACA 34 gel filtration
6 Chromatofocusing
7 Ultrogel ACA 34 gel filtration
8 Zn chelating Sepharose 6B chromatography
9 Fast protein liquid chromatography

Material and Methods

Renin Substrate Purification

Renin substrate was purified from pooled plasma of healthy nonpregnant or pregnant subjects. The purification is described in detail in table I. Briefly, pooled Na$_2$EDTA plasma was delipidized with silicate, followed by centrifugation. The fraction containing RS was precipitated with 2.3 *M* (NH$_4$)$_2$SO$_4$. Following dialysis, DEAE-sepharose 6B (Pharmacia Fine Chemicals, Uppsala, Sweden) ion exchange chromatography was performed. The pooled and dialyzed RS peak was then further purified using gel filtration on an ACA 34 column (LKB, Bromma, Sweden), chromatofocusing using polybuffer 74 and polybuffer exchanger 94 (Pharmacia Fine Chemicals), and then gel filtration on Ultrogel ACA 34 to remove buffer components. This preparation yielded one main band on SDS gel electrophoresis, corresponding to 56,000 daltons, the approximate molecular weight of RS. To achieve nearly complete purity, the preparation was chromatographed on a Mono Q column using a fast protein liquid chromatography (FPLC) apparatus (Pharmacia Fine Chemicals). After sample injection in 20 mmol/l Tris-HCl buffer pH 8 the sample was eluted with a linear gradient of NaCl from 0 to 500 mmol/l in the same buffer using a flow rate of 2 ml/min.

Renin Substrate Assay

RS was measured either using indirect assay of angiotensin I, generated during incubation with human kidney renin [6], or using a direct RIA method [Metsärinne et al., to be published]. Antisera against RS were raised in rabbits by monthly subcutaneous and intradermal multiple-site injections of purified RS with Freund's complete adjuvant (first injection) or incomplete (booster doses) adjuvant (Difco Laboratories, Detroit, Mich.). RS was iodinated with the chloramine T method.

EP Antisera and Hemagglutination Reagents

Reagents for hemagglutination inhibition assay of human EP, and for agglutination studies, were from JCL Clinical Research Corp., Knoxville, Tenn. Antisera against human urinary EP were kind gifts of Dr. J.W. Fisher, (New Orleans), and Dr. E.D. Zanjani, (Minneapolis, Minn.). Both antisera are capable of blocking the biological action of EP.

Purified human RS was coupled to washed human 0 Rh+ erythrocytes by glutaraldehyde condensation [Rosenlöf, to be published].

RS and EP Staining of Cerebellar Hemangioblastoma and Renal Tumors

Eight tumors from patients with cerebellar hemangioblastoma associated with high (n = 4) or high normal (n = 4) hemoglobin values were stained as described elsewhere [7] for IFL studies with antisera against RS, EP and control antisera. Likewise, 8 tumors from patients with renal carcinoma, 4 of which were associated with secondary erythrocytosis, were similarly stained.

Results

Purification of Renin Substrate

RS purified from human pregnancy plasma showed one major band on SDS-gel electrophoresis (fig. 1) corresponding to about 56,000 daltons. The main purification steps, leading to the purest preparation, which yielded 24 μg angiotensin I upon incubation with human kidney renin, are indicated in table I. Following FPLC, purified nonpregnant RS differed from RS prepared from pregnancy plasma. In pregnancy plasma, the major peak of protein corresponded to RS as measured both with direct and indirect RS assay methods, while nonpregnant plasma in addition to that peak contained a major peak which did not yield angiotensin I, nor did it react in the direct RIA.

Cross-Reaction of RS with EP

As reported earlier [4], antisera against human RS showed strong cross-reaction with purified human urinary erythropoietin by agglutinating erythrocytes covered with glutaraldehyde-coupled erythropoietin. Control sera were consistently negative.

When erythrocytes covered with purified RS were incubated with two different antisera against purified human urinary EP (Fisher, Zanjani), strong and reproducible agglutination was seen, as will be reported elsewhere [Rosenlöf, to be published]. Interestingly, the capacity of EP antisera to agglutinate RS-covered cells correlated well to the reported potency of these antisera to block the biological action of EP. Control antisera and rabbit serum were negative.

RS and EP Staining of Tumor Cells

All cerebellar hemangioblastoma tumors contained cells of roughly similar morphological appearance, located in the stroma, which stained strongly positively with both antiserum against human EP (Fisher) and our antiserum against RS (fig. 2). These cells also stained positively with antiserum against human alpha$_1$-antitrypsin, but stained negative with control rabbit serum, antiserum against glial filaments, human factor VIII, and laminin [5].

All renal carcinoma tumors stained negative with both the EP and RS antisera tested.

Discussion

The present data form a basis for a critical analysis of our hypothesis of the possibility that RS may be an EP precursor (fig. 3). The preparations of human RS used in these studies, have repeatedly shown cross-reaction with available preparations of human erythropoietin. Further, we show here that the fraction purified on HPLC, corresponding to RS with direct RIA method also yielded angiotensin I upon incubation with renin.

The demonstration of strong agglutination of RS-covered erythrocytes with two well-characterized antisera against human EP, made in two different laboratories further emphasizes the antigenic kinship of RS and EP. Such a relation is difficult to conceive if one does not accept the concept that RS and EP are closely related glycoproteins. We have presented preliminary, albeit not conclusive data, that RS may be an erythropoietin precursor [4]. This idea is further corroborated by the present results. However, a recent report on the cloning of the gene coding for fetal human erythropoietin [8] apparently casts some doubt on this hypothesis. Thus, these workers could find no structural homology between rat angiotensinogen and the amino acid sequence of human fetal liver EP deduced by them. This might of course imply that RS is not an EP precursor. It must be emphasized that human angiotensinogen shows no cross-reaction with antisera against rat angiotensinogen, however [9]. Another explanation for this discrepancy could be that there may in fact exist different erythropoietins, e.g. a fetal one and an adult molecule, not necessarily closely related. Several recent reports on various erythropoiesis-stimulating factors in human blood would

Fig. 1. Sodium dodecyl sulphate (SDS) gel electrophoresis of purified human renin substrate (lane 1, and control substances). Lane 2 = Ovalbumin; lane 3 = hemoglobin A; lane 4 = human serum albumin; lane 5 = transferrin. Numbers refer to molecular weight in daltons.

Fig. 2. Renin substrate immunoreactive cells in cerebellar hemangioblastoma tumor, associated with secondary erythrocytosis. Immunofluorescence labeling. × 1,310.

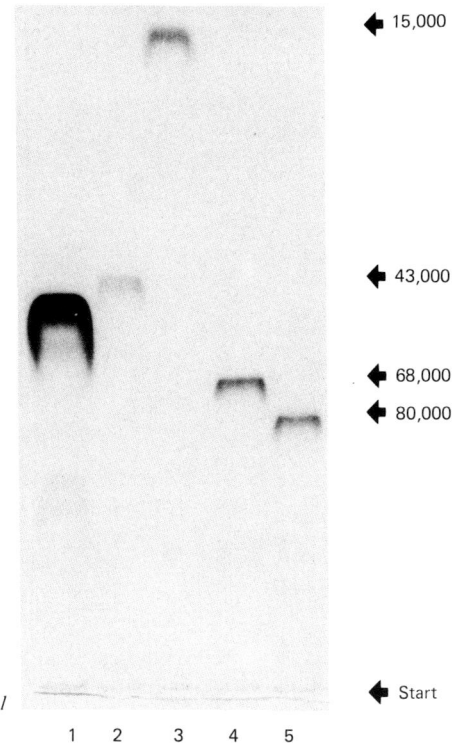

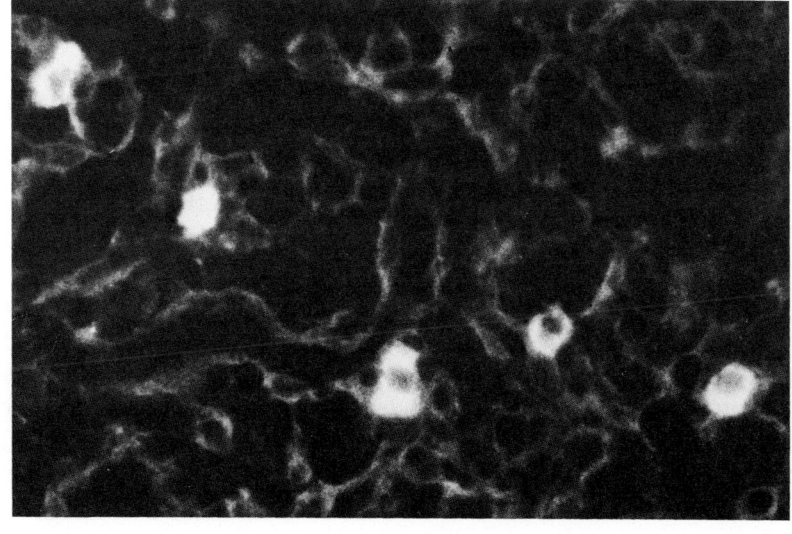

Present hypothesis

Renin substrate (angiotensinogen) is a precursor of erythropoietin

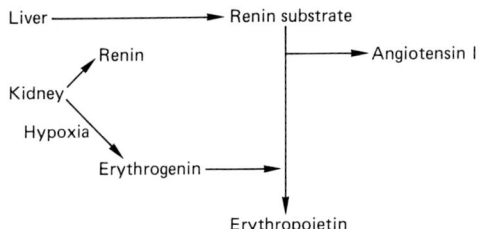

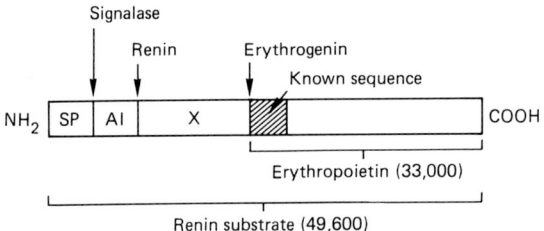

Fig. 3. View of the hypothesis. The upper panel shows the hypothetical origin of ery-thropoietin from renin substrate. The lower panel shows the molecular basis for the hypothesis. NH$_2$ = Amino-terminal end; SP = signal peptide; AI = angiotensin I; erythroge-nin = hypothetical kidney enzyme; X = unknown amino acid sequence, cleaved off by ery-throgenin.

fit such a polymorphic view of erythropoietin(s). In any case, our data show a strong immunological relation between EP and RS, which warrants further clarification.

The coexistence in cerebellar hemangioblastoma cells of both EP and RS, as reported here, is of considerable interest in this context, and pro-vides clinically based support to our hypothesis on a relation between RS and EP. Thus, these tumors are known to be associated with secondary ery-throcytosis, which was the case in at least 4 of the 8 patients reported here, and is to be reported in detail elsewhere [5]. Such tumors have been claimed to contain bioactive EP before, while we are aware of no other reports de-monstrating both EP and RS in cerebellar hemangioblastoma. Their coexis-tence in this tumor would be conveniently explained by RS being an EP pre-cursor. This tempting explanation, of course, is far from being proven as yet.

References

1   Jacobsen, L.O.; Goldwasser, E.; Friend, W.; Plzak, L.: Role of the kidney in the ery-
    thropoiesis. Nature, Lond. *179:* 633–634 (1957).
2   Kuratowska, Z.; Lewartowski, B.; Michalak, E.: Studies on the production of erythro-
    poietin by the isolated perfused organs. Blood *18:* 527–534 (1961).
3   Gould, A.B.; Goodman, S.; DeWolf, R.; Onesti, G.; Swartz, C.: Interrelation of the
    renin system and erythropoietin in rats. J. Lab. clin. Med. *96:* 523–534 (1980).
4   Fyhrquist, F.; Rosenlöf, K.; Grönhagen-Riska, C.; Hortling, L.; Tikkanen, I.: Is renin
    substrate an erythropoietin precursor? Nature, Lond. *308:* 649–652 (1984).
5   Fyhrquist, F.; Rosenlöf, K.; Grönhagen-Riska, C.; Böhling, T.; Haltia, M.: Erythro-
    poietin and renin substrate in cerebellar haemangioblastoma. Acta med. scand. (in
    press, 1985).
6   Immonen, I.; Fyhrquist, F.; Pohjavuori, M.; Simell, O.: Age dependence of human
    plasma renin substrate. Scand. J. clin. Lab. Invest. *41:* 167–170 (1981).
7   Böhling, T.; Paetau, A.; Ekblom, P.; Haltia, M.: Distribution of endothelial and base
    membrane markers in angiogenic tumours of the nervous system. Acta neuropath. *62:*
    67–72 (1983).
8   Jacobs, K.; Shoemaker, C.; Rudersdorf, R.; et al.: Isolation and characterization of
    genomic and cDNA clones of human erythropoietin. Nature, Lond. *313:* 806–810
    (1985).
9   Bouhnik, J.; Clauser, E.; Gardes, J.; Corvol, P.; Menard, J.: Direct radioimmunoassay
    of rat angiotensinogen and its application to rats in various endocrine states. Clin. Sci. *62:*
    355–360 (1982).

Prof. F. Fyhrquist, MD, Helsinki University Central Hospital,
IVth Department of Medicine, Unioninkatu 38, SF–00170 Helsinki 17 (Finland)

Discussion

Koch: Thank you for your very beautiful presentation. Questions?

Drueke: There is another model of in vitro increased production of erythropoietin in mouse leukemia cells. Do you know whether, in this model, there is also increased angiotensinogen production, or staining in those cells, together with erythropoietin?

Fyhrquist: I am sorry, I do not know. I would like to try that.

Forssmann: I would like to ask you: Is there really no homology in the sequence which you have in your graph – the part which has been sequenced in the cDNA? Is that what you compared?

Fyhrquist: There is no clear homology between the molecular sequence – amino acid sequence – reported by Jakobs et al. [Nature, Lond. *313:* 806–810 1985], I referred to. The molecule reported for rat angiotensinogen by Nakanishi and co-workers has appeared in Proc. natn. Acad. Sci. USA *80:* 2196–2200 (1983).

Kurtz: Two questions please. The first is, have you ever tried to activate angiotensinogen with a renal protease? The second is, would you agree, too, that from your findings one should be very careful with the interpretation of radioimmunoassay data about erythropoietin because if you make a RIA for erythropoietin, you have shown that angiotensin at least will also react in this RIA and you cannot talk about the real role of erythropoietin.

Fyhrquist: In response to your second question, I can only say that I heartly agree with you. Yes, one should be very careful in the interpretation of the immunoassay data. We have excluded the possibility of cross-reaction. To your first question, yes, we have tried but they are so preliminary that I would rather not tell about them. That is an obvious experiment to do: trying to find a protease which would generate erythropoietin.

Forssmann: I would again like to ask you how many amino acid residues have been sequenced in your investigations?

Fyhrquist: We have not sequenced our preparation, what the figure 1 showed was presenting a hypothesis. The sequencing of our preparation is about to start.

Ritz: It was not perfectly clear about this experiment with the erythrocyte leukemia cell where you induced benzidine-positive and Hb-positive cells. Were you culturing these cells in medium free of fetal calf serum or was fetal calf serum added.

Fyhrquist: Yes, it was fetal calf serum.

Ritz: This is, of course, a serious variable to be taken into account.

Fyhrquist: Of course. It is possible that calf serum contained some factor, which, together with the renin substrate preparation, caused induction of K562 cells. We are fully aware of that. Fetal calf serum alone did not induce HbF synthesis in those cells. It needed our preparation of renin substrate.

Koch: If you put your data and Dr. Kurtz' data together, would you conclude that there have to be other precursors than renin substrate to erythropoietin?

Fyhrquist: Yes, there is an additional possibility that in fact we have several erythropoietins. It might be a polymorphic hormone.

Contr. Nephrol., vol. 50, pp. 175–187 (Karger, Basel 1986)

Erythropoietin Production in Cultures of Rat Renal Mesangial Cells

Armin Kurtz[a], Wolfgang Jelkmann[b], Josef Pfeilschifter[a], Christian Bauer[a]

[a]Institute of Physiology, University of Zürich, Zürich, Switzerland;
[b]Institute of Physiology, Medizinische Hochschule, Lübeck, FRG

It is a well-known phenomenon for clinicians as well as for physiologists that states of insufficient oxygen delivery to the organism are normally followed by an increase in red cell mass. This rise in erythropoiesis is evoked by an increased plasma concentration of the glycoprotein hormone erythropoietin. Under extreme situations 100-fold increases in the plasma concentration of erythropoietin in response to hypoxia can be observed in man as well as in laboratory animals [1]. Moreover, it is known that about 90% of the occurring erythropoietin is generated by the kidneys during hypoxic situations [2]. From these findings it can be concluded that the kidney contains a kind of oxygen sensor that induces formation of erythropoietin during insufficient oxygen supply to the kidney. The localization and working mechanism of this oxygen sensor has attracted considerable interest among different research groups including our own. Since about 3 years the main part of our research is concentrated on the question how this oxygen sensor works, i.e. how the signal hypoxia is translated into enhanced erythropoietin formation. An investigation of this problem by the use of the whole kidney seemed to us to be less successful because of the complexity of the organ system. Cell cultures of the renal erythropoietin-producing cells, however, appeared to be a more suitable experimental model and we therefore decided to culture erythropoietin-producing renal cells. Which cell, however, generates erythropoietin in the kidney? Investigations done in the 1960s [3] and 1970s [4] using erythropoietin antisera in combination with indirect immunofluorescence staining revealed that only renal glomeruli showed a positive and significant staining in the kidneys of sheep and rats exposed to

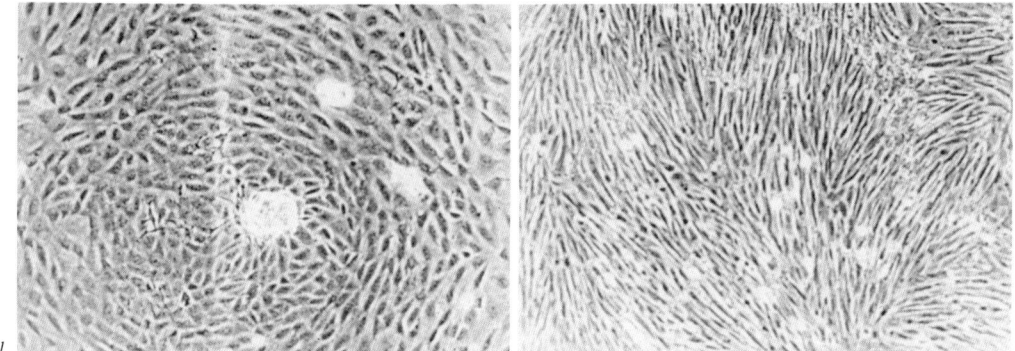

Fig. 1. Phase-contrast photomicrograph of a primary culture of rat renal glomeruli 1 week after onset of culture. × 116.

Fig. 2. Phase-contrast photomicrograph of a primary culture of rat renal glomeruli 4 weeks after onset of culture. × 116.

hypoxia. In view of these results we therefore decided to culture rat renal glomeruli and to investigate whether or not a certain glomerular cell type is capable of producing erythropoietin.

Glomerular Cell Culture

The glomerulus without juxtaglomerular apparatus normally consists of four cell types, namely the epithelium of the Bowman capsule, the podocytes, the capillary endothelium and the mesangial cells in the core of the glomerulus. For isolation of the glomeruli from the kidneys [5] we used a simple sieving technique first described by Krakower and Greenspon [6]. Using this technique, the glomeruli are detached from the renal tissue at the vascular pole. About 20% of the isolated glomeruli still have the Bowman capsule and such a glomerulus preparation then theoretically contains four cell types, namely endothelial cells, podocytes, mesangial cells and epithelial cells from the Bowman capsule. Contamination of the glomerulus preparation with tubular fragments and single cells is less than 2%, as confirmed by phase-contrast microscopy. When these glomeruli are cultured under cell culture conditions, the glomeruli normally attach to the surface of the culture dish during the first 2 days. With the third day of culture the first

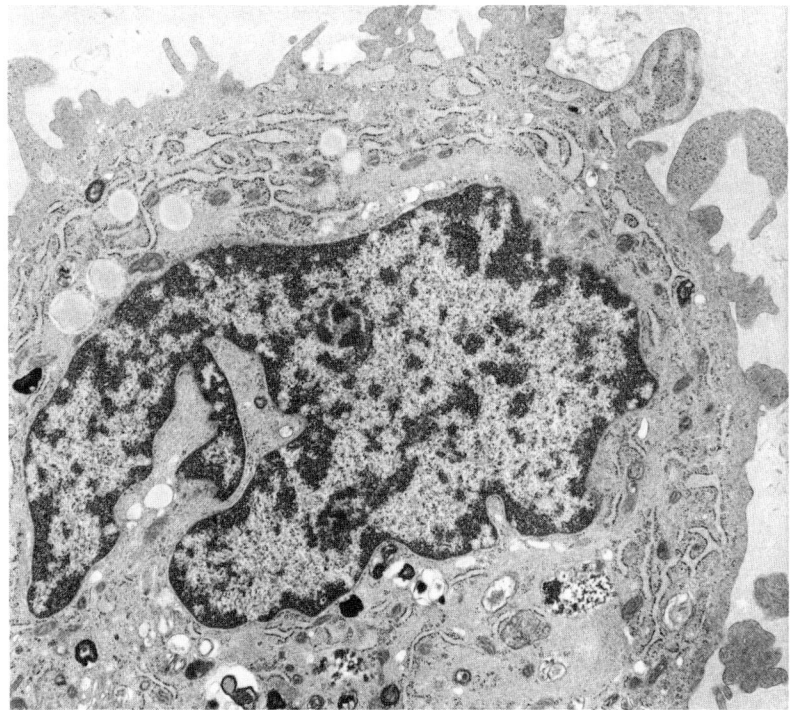

Fig. 3. Transmission electron microscopy of a cultured mesangial cell. × 9,800.

cells begin to grow out from the attached glomeruli. After the first week a typical cell culture is obtained, as can be seen in figure 1. What I shall say about the characterization of the cultured glomerular cells in the following is mainly based upon the results obtained by a number of research groups who are concerned with the identification of cultured glomerular cells [for review, see 7]. The cells which dominate the culture after the first week (fig. 1) grow in a contact-inhibited manner. According to a board of characteristics such as electron-microscopical ultrastructure [8], thymidine incorporation kinetics [9], and other characteristics, these cells are believed to be epithelial cells. After 2–3 weeks of the primary culture a drastic change in the culture characteristics occurs. The cobblestone-like epithelial cells perish and they are replaced by cells which grow in an overlapping manner and have a spindle-shaped form (fig. 2). These cells are believed to be mesangial cells [7]. This conclusion is confirmed by a number of findings: these cells contract upon addition of angiotensin II and arginine vasopressin [10] as is known from mesangial cells in situ. Our own investigations revealed that these cells slow down their growth rate upon addition of D-valine-containing medium but do not detach from the culture dish during 14 days with repeated exchange of the culture medium. A detachment of cells from the surface upon addition of D-valine-containing medium would be typical for fibroblasts, in particular fibroblasts from rodents [11]. Figure 3 shows an electron-microscopic photograph of these cultured cells and this picture strongly resembles mesangial cells in situ. Please note the indented nucleus and the high number of microfilaments surrounding the nucleus. Our strongest argument, however, that these cells are mesangial cells is our finding that these cells contain a cytoskeleton, as is typical for mesangial cells in situ. With the help of Prof. Weber from the Max-Planck Institute in Göttingen we found that these cultured cells contain the intermediate filaments vimentin and desmin. Control experiments with the kidney showed that within the glomerulus only the mesangial cells contain desmin and vimentin [12]. Still, the existence of endothelial cells in the culture has to be considered. To our knowledge, however, only one group has been successful in culturing endothelial cells from renal glomeruli [3]. All other investigators including us did not obtain any evidence for a growth of endothelial cells under these culture conditions.

Erythropoietin Production by Cultured Glomerular Cells

To find out whether glomerular cell cultures produce erythropoietin or not, we have tested the culture medium for erythropoietin activity during the time course of the culture [5]. For the determination of erythropoietin activity we used the fetal mouse liver cell assay for erythropoietin [14] and polycythemic mouse assay for erythropoietin [15]. We failed to detect any activity during the first week of culture, even when the glomeruli were seeded with high density so that the epithelial cells in the culture already formed a monolayer after 1 week. To the degree, however, to that mesangial cells developed in the culture, erythropoietin activity occurred which was also demonstrable in passages of mesangial cell cultures. We then characterized this erythropoietin activity according to its biological, physical and chemical characteristics and we hereby found no differences to erythropoietin. Therefore, we conclude from these results that mesangial cells in culture are capable of producing erythropoietin and we furthermore con-

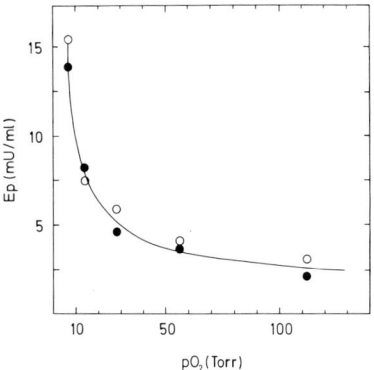

Fig. 4. Relationship between erythropoietin activity in the culture medium and the partial pressure of oxygen in the incubation atmosphere. Two different tests under same conditions.

clude that mesangial cells are a likely candidate for erythropoietin formation in the kidney.

Our interest then concentrated on the question whether or not mesangial cells in culture also possess an oxygen sensor comparable with the situation in vivo. Up to this we determined the erythropoietin production rate of cultured mesangial cells at different oxygen concentrations in the incubation atmosphere (fig. 4).

For these experiments, cultured mesangial cells were incubated at the indicated oxygen concentrations in the incubation atmosphere. Figure 4 shows the erythropoietin concentration in the culture medium after a 2-day incubation. The erythropoietin activity in the culture medium was indeed dependent on the partial pressure of oxygen. It is obvious that the erythropoietin formation rate of the cultured mesangial cells was inversely related to the oxygen pressure. We conclude from this finding that cultured mesangial cells contain an oxygen sensor which translates the signal hypoxia in enhanced erythropoietin formation. Looking for the mechanisms by which this sensor could work, we obtained an interesting result from in vivo experiments. From investigations done by the group of Fisher [16], it was known that the stimulation of erythropoietin production by hypoxia could be prevented by inhibition of the prostaglandin system. We repeated these experiments and we further investigated the effect of the cyclooxygenase inhibitor indomethacin on the in vivo erythropoietin formation of rats during hypoxia, phenylhydrazine anemia and cobalt treatment [17]. We found that

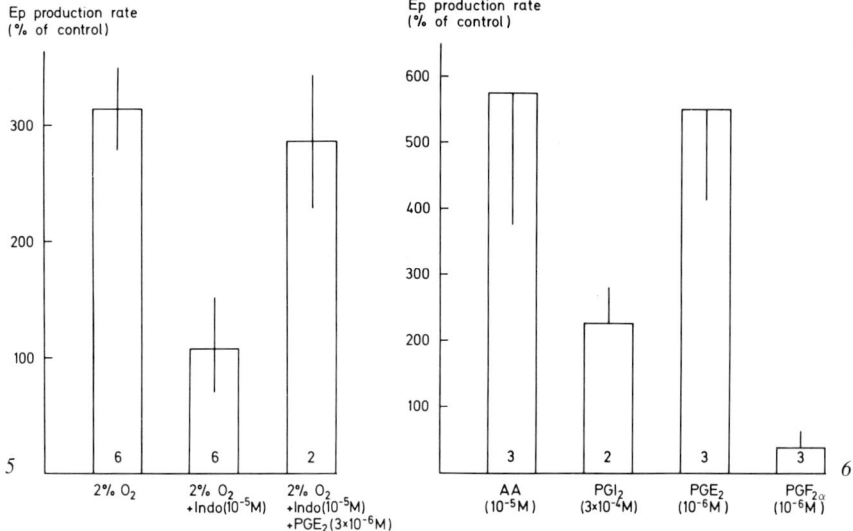

Fig. 5. Effects of indomethacin, alone and with PGE_2 on the erythropoietin production rate of cultured mesangial cells at 2% O_2. Ep production rate is given as percent of control (20% O_2). Data are presented as mean ± SEM. The figures at the bottom of the columns indicate the number of independent experiments.

Fig. 6. Effects of added arachidonic acid (AA), PGI_2, PGE_2 and $PGF_{2\alpha}$ on the Ep production rate of cultured mesangial cells at 20% O_2. Values and data are presented as described in the legend to figure 5.

the stimulation of erythropoietin formation by hypoxia and anemia but not that by cobalt could be prevented by indomethacin. This finding could be an indication that prostaglandins might have a possible role in the translation mechanism between hypoxia and enhanced erythropoietin formation. We therefore investigated the role of prostaglandins during erythropoietin formation in our cultured mesangial cells. We found that the stimulation of erythropoietin formation by hypoxia could be inhibited by the cyclooxygenase inhibitor indomethacin (fig. 5). This inhibitory effect of indomethacin could be abolished by simultaneously adding prostaglandin E_2.

If the prostaglandins have a causal and not only permissive role in hypoxia-stimulated erythropoietin formation then they should also be able to stimulate erythropoietin formation independently of hypoxia. Figure 6 shows the effects of the prostaglandin precursor arachidonic acid and of prostacyclin, prostaglandin E_2 and prostaglandin $F_{2\alpha}$ on erythropoietin for-

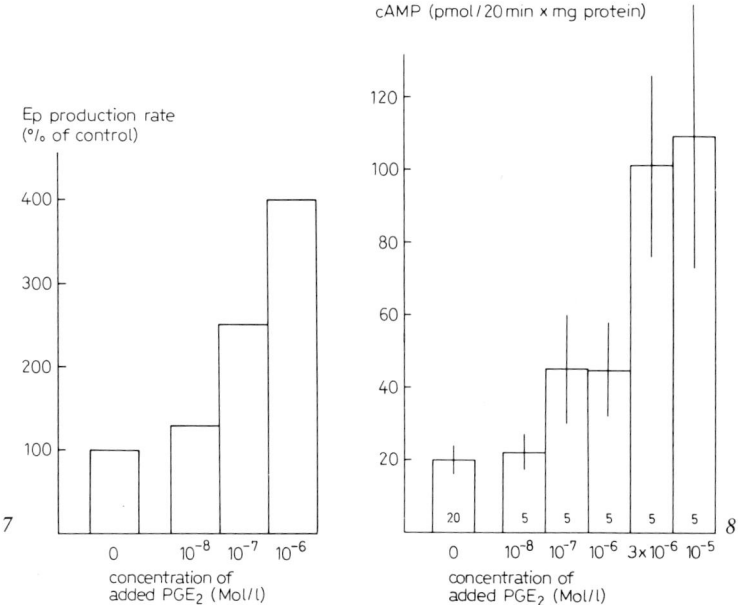

Fig. 7. Effects of different concentrations of added PGE_2 on the Ep production rate of cultured mesangial cells at 20% O_2. The experiments were performed simultaneously using one culture line. Data are presented as described in the legend to figure 5.

Fig. 8. Intracellular cAMP formation in cultured mesangial cells with different concentrations of added PGE_2. The figures at the bottom of the columns indicate the number of independent experiments.

mation at normal (20%) oxygen concentrations in the incubation atmosphere. It is evident that arachidonic acid, prostacyclin and prostaglandin E_2 stimulate erythropoietin formation whilst $PGF_{2\alpha}$ rather seems to have an inhibitory effect. If prostaglandins are signal molecules for the mediation of hypoxia to enhanced erythropoietin formation, then an increased prostaglandin formation during hypoxia should be expected. And indeed we found that hypoxia increased prostaglandin formation in the cultured mesangial cells by a factor of about 2 [18]. Interestingly, the pattern of prostaglandins was not changed by hypoxia. The main prostaglandin was prostaglandin E_2 followed by prostaglandin $F_{2\alpha}$ and prostacyclin. From this finding it can be concluded that during hypoxia an increased release of arachidonic acid, which is the rate-limiting substrate for prostaglandin formation has occurred. At present we are looking for the mechanism for enhanced release of arachidonic acid during hypoxia. Besides this point we are also concerned

Table I. Erythropoietin production and intracellular cAMP formation by cultured mesangial cells in presence and absence of forskolin (data are mean ± SEM)

	Erythropoietin production rate mU/flask × day	cAMP formation pmol/mg protein × min
Control	16 ± 5 (n = 4)	20 ± 4 (n = 4)
Forskolin ($10^{-5} M$)	45 ± 6 (n = 3)	789 ± 128 (n = 4)

with the question how prostaglandins might influence erythropoietin formation. The finding that prostacyclin and prostaglandin E_2, which are well-known activators of the adenylate cyclase in many cell types, but not prostaglandin $F_{2\alpha}$ stimulated erythropoietin formation could suggest that the effect of prostaglandins on erythropoietin formation is mediated by an activation of adenylate cyclase.

Figure 7 shows, for PGE_2 as an example, the dose dependency of erythropoietin formation on the concentration of prostaglandin E_2. Figure 8 shows the dose dependency of the adenylate cyclase activity of culture mesangial cells on the prostaglandin E_2 concentration. It is obvious that the dose-response curves for erythropoietin formation and adenylate cyclase formation are very similar. We furthermore found that prostacyclin and arachidonic acid which were found to stimulate erythropoietin formation of cultured mesangial cells also stimulated adenylate cyclase activity in the cultured cells. If the activation of the adenylate cyclase is causally involved in the mediation of the prostaglandin effect on erythropoietin formation, then the prostaglandin-independent stimulation of adenylate cyclase should also stimulate erythropoietin formation. We therefore investigated the effects of a direct stimulation of adenylate cyclase by forskolin on erythropoietin formation in the cultured mesangial cells. As can be seen in table I, we found that forskolin $10^{-5} M$ increased both intracellular cAMP and erythropoietin formation. The stimulation of erythropoietin formation by forskolin was exactly in the range of magnitude of the effects of prostaglandins and hypoxia on erythropoietin formation. The results obtained with forskolin therefore led to the conclusion that prostaglandins are likely to enhance erythropoietin formation in cultured mesangial cells via an activation of adenylate cyclase and herein by its product cAMP.

Our results are compatible with the conclusion that renal mesangial cells are a site of erythropoietin formation in the kidney. Cultured mesangial cells contain an oxygen sensor which translates hypoxia in enhanced erythropoietin formation. In this sensor mechanism prostaglandin E_2 and/or prostacyclin are likely to be involved, which increase during hypoxia. These prostaglandins stimulate the formation of erythropoietin via an increase in the intracellular concentration of cAMP, which is caused by a prostaglandin-dependent stimulation of adenylate cyclase.

References

1　Erslev, A.J.; Caro, J.; Miller, O.; Silver, R.: Plasma erythropoietin in health and disease. Ann. clin. Lab. Sci. *10:* 250−257 (1980).

2　Erslev, A.J.; Caro, J.; Kansu, E.; Silver, R.: Renal and extrarenal erythropoietin production in anaemic rats. Br. J. Haemat. *45:* 65−72 (1980).

3　Fisher, J.W.; Taylor, G.; Porteous, D.D.: Localization of erythropoietin in the glomeruli of sheep kidney using a fluorescent antibody technique. Nature, Lond. *205:* 611−612 (1965).

4　Gruber, D.F.; Zucali, J.R.; Wleklinski, J.; La Russa, V.; Mirand, E.A.: Temporal transition in the site of rat erythropoietin production. Exp. Hematol., Copenh. *5:* 399−407 (1977).

5　Kurtz, A.; Jelkmann, W.; Bauer, C.: Mesangial cells derived from rat glomeruli produce an erythropoiesis stimulating factor in cell culture. FEBS Lett. *137:* 129−132 (1982).

6　Krakower, C.A.; Greenspon, S.A.: Localization of the nephrotoxic antigen within isolated renal glomerulus. Archs Path. *51:* 629−639 (1954).

7　Kreisberg, J.I.; Karnovsky, M.J.: Glomerular cells in culture. Kidney int. *23:* 439−447 (1983).

8　Kreisberg, J.I.; Hoover, R.L.; Karnovsky, M.J.: Isolation and characterization of rat glomerular epithelial cells in vitro. Kidney int. *14:* 21−30 (1978).

9　Nörgaard, J.O.R.: Cellular outgrowth from isolated glomeruli. Lab. Invest. *48:* 526−542 (1983).

10　Ausiello, D.A.; Kreisberg, J.I.; Roy, C.; Karnovsky, M.J.: Contraction of cultured rat glomerular cells of apparent mesangial origin after stimulation with angiotensin II and arginine vasopressin. J. clin. Invest. *65:* 754−760 (1980).

11　Gilbert, S.F.; Migeon, B.R.: *D*-Valine as a selective agent for normal human and rodent epithelial cells in culture. Cell *5:* 11−17 (1975).

12　Bachmann, S.; Kriz, W.; Kulm, C.; Franke, W.W.: Differentiation of cell types in the mammalian kidney using antibodies to intermediate filament proteins and desmoplakins. Histochemistry *77:* 365−394 (1983).

13　Striker, G.E.; Soderland, C.; Bowen-Pope, D.F.; Gown, A.M.; Schmer, G.; Johnson, A.; Luchtel, D.; Ross, R.; Striker, L.J.: Isolation, characterization and propagation in vitro of human glomerular endothelial cells. J. exp. Med. *160:* 323−328 (1984).

14   Kurtz, A.; Jelkmann, W.; Bauer, C.: Insulin stimulates erythroid colony formation inde-
      pendently of erythropoietin. Br. J. Haemat. *53:* 311−316 (1983).
15   Jelkmann, W.; Bauer, C.: Demonstration of high levels of erythropoietin in rat kidneys
      following hypoxic hypoxia. Pflügers Arch. *392:* 34−39 (1981).
16   Fisher, J.W.: Prostaglandins and kidney erythropoietin production. Nephron *25:* 53−56
      (1980).
17   Jelkmann, W.; Kurtz, A.; Seidl, J.; Bauer, C.: Mechanisms of the renal glomerular
      erythropoietin production; in Grote, Witzleb, Erwin Riesch Symp., 1984, pp. 130−137
      (Erich Steiner Verlag, Stuttgart 1984).
18   Jelkmann, W.; Kurtz, A.; Förstermann, U.; Pfeilschifter, J.; Bauer, C.: Hypoxia en-
      hances prostaglandin synthesis in renal mesangial cell cultures. Prostaglandins *30:*
      109−118 (1985).

Dr. A. Kurtz, Physiologisches Institut der Universität, Winterthurerstrasse 190,
CH−8057 Zürich (Switzerland)

Discussion

Koch: Thank you for your very exciting presentation. Questions?

Massry: These are really beautiful data. May I ask you, what do you suggest by which
mechanism hypoxia stimulates prostaglandin? You have heard Dr. Kirschenbaum talking
about ionophor stimulating prostaglandins as a consequence of change in cytosolic calcium.
Now you know that cytosolic calcium is controlled by about six or seven pumps within the
membrane or within the intracellular structures, and it is not unlikely that hypoxia affects the
function of these pumps and allows accumulation of calcium within the cell. I think it would be
very nice, if you can, to measure cytosolic calcium and that will probably provide you with the
link.

Kurtz: Yes, we are planning to do this by using two methods. Our thought is that perhaps
phospholipase A could be activated and this is very hot data. We are just doing the experiments. We find a breakdown of polyphosphoinositides − and this would indicate that during
hypoxia a phospholipase is indeed activated. We do not know the activator. We do not know
whether a lack of oxygen can activate a phospholipase. It might be that it is a metabolic substrate that, in turn, activates the phospholipase.

Massry: I must also mention to you, if you plan to measure intracellular calcium that
these cells are adhered to the plate. Now, scraping them, to make them free for the measurement will result in a tremendous leak of the intracellular calcium stores because you are going
to affect their integrity and I think you should be very careful in calculating and providing for
the leak, otherwise you are going to get false data.

Kurtz: We did a lot of studies of prostaglandin synthesis by these cultured mesangial cells
in combination with the vasoconstrictors Dr. Kirschenbaum mentioned before with angiotensin II AVP and norepinephrine and we found that each stimulation of prostaglandin synthesis
requires some calcium influx into the mesangial cells. So, it might be that there is also an increase in flux during hypoxia. Our next experiment is designed to prevent this influx by calcium

channel blocker. So, if we can prevent the rise in prostaglandin by calcium channel blockers, this would be a strong argument for an increase of calcium influx during hypoxia.

Ritz: I would appreciate it if you could comment on two points. The first – there has been some controversy about the dialysation of circulating erythropoetin. Could you please comment to what extent the erythropoetin that is generated in glomerular cell cultures corresponds to the circulating form? Second, in a more general way I think that the evidence that you presented for an interaction between prostaglandin system and erythropoetin release is quite convincing. Still, one is puzzled why treatment with cyclooxygenase inhibitors in patients or animals does not cause anemia. Could you explain the paradox?

Kurtz: In response to your first question I cannot give an exact answer for the ratio of sialated and desialated erythropoietin. Our first experience with erythropoietin from mesangial cells was that we only got in our culture the desialated erythropoietin. Then a paper came to our notice in which the investigators had found that the composition of the culture medium, for example Hepes, can radically alter the glycosylation of glycoproteins. We then took Hepes out of our culture medium and got an in vivo activity of erythropoietin. This in vivo activity was comparable with the activity of the desialated erythropoietin. So we always have to consider an culture artefact when we consider sialated and desialated erythropoietin.

Ritz: Why doesn't one observe anemia with indomethacin?

Kurtz: We don't know the role of prostaglandin in the basic erythropoietin production. If we add indomethacin to cells kept at 20% oxygen, we get no decrease in erythropoietin formation. We can only prevent the hypoxia-induced stimulation.

Ritz: So the experiment to do would be to treat an animal or patient with indomethacin, while he or it is exposed to hypoxic challenge.

Kurtz: Yes, that would be a very good experiment.

Drueke: If the capacity of the mesangial cells to produce erythropoietin is maintained, for how much time and whether from one culture to the other do you get the same amount of erythropoietin produced in response to stimulators?

Kurtz: We investigated erythropoietin formation till the fourth passage and that means that the cells had an age of about 60 days. Until this stage, we got an erythropoietin formation. It turned out that erythropoietin formation was best in the first passage and then decreased.

Gross: In your experiment, you had to use hypoxia of about 13 mm Hg to produce erythropoietin release and this was in the cell-free suspension which makes it a lot worse as compared with the perfused glomerulus in vivo. To reach a lower partial pressure of oxygen should be extremely difficult.

Kurtz: The values I have shown are related to the oxygen concentration in the incubator. The cells are incubated for 2 days at this oxygen concentration. Of course, if you superfuse material with an oxygenated solution, you always have leaks from which oxygen will diffuse, but if you have it in an incubator which is tightly closed you have no problems with leakage.

Gross: This is still much less than ever reached physiologically.

Kurtz: There are problems. We do not know the exact partial pressure of oxygen at the mesangial cells.

Kirschenbaum: I also enjoyed your presentation very much. When I first got into prostaglandin research 13 or 14 years ago, I remember a reading that hypoxia was a general stimulus for prostaglandin production by cells in general and this may not be a specific phenomenon for these mesangial cells obviously, but the general membrane phenomenon too, perhaps, increases phospholipase activity as such. An explanation, perhaps, for Dr. Ritz' question on why nonsteroidals do not decrease erythropoietin levels may be more complicated perhaps than your explanation. The levels of prostaglandins produced by remaining renal tissue may be much in excess, as I have demonstrated, of levels that exist under basal conditions, and the concentrations of indomethacin and other nonsteroidals that can be administered to patients do not approach the levels which we can use in vitro. So the net effect may be that there is still excess of amounts of prostaglandins present at the level of the mesangial cell if that is the site of presentation, that the changes that we can see in vivo may not be comparable to those that can be obtained in vitro. That leads me to my final question. The doses of PGE_2 that you use were pharmacologic doses in this system. It looked like your lowest concentration was 10^{-6} or 10^{-7} mol?

Kurtz: It was the highest concentration, and the lowest was 10^{-8} mol. This is the exact concentration we observed in the culture after exchange of the culture medium of 2 days. So we did not go to lower concentrations.

Eigler: Have you any evidence that the mesangial cells may be the major site of total erythropoietin production under normal conditions?

Kurtz: No, we have not done experiments to this point. It is known that the liver is capable of producing erythropoietin. Perhaps also the bone marrow is capable of producing erythropoietin.

Eigler: I mean within the kidney, of course.

Kurtz: We have not investigated this, but there is a study done by Erslev who tried to localize erythropoietin in the kidney and he made a glomerular fraction and a tubular fraction. There he found more erythropoietin in the tubular than in the glomerular fraction and from that he concluded that the tubules should be the formation site of erythropoietin. I think he made one mistake in his calculation. The glomeruli account for about 2% of the whole kidney mass and tubules account for about 90%, and if you relate it to the protein, what he did not, you get a specific activity of erythropoietin which is 10 times higher in the glomerular than in the tubular fraction. There is no doubt that erythropoietin is filtrated, otherwise you would not get such high increases in erythropoietin in the urine of patients with aplastic anemia. There is also no doubt that erythropoietin is reabsorbed by the tubules.

Heidland: You mentioned the involvement of adenylate cyclase in the erythropoietin production. Can this activation be inhibited by beta-blocking agents?

Kurtz: We did not do this experiment. We did another experiment with stimulation by beta-agonists. Under this condition you also get an in vivo active erythropoietin.

Ritz: Let me come back to Dr. Eigler's question with respect to localisation of erythropoietin in vivo. Of course, one would require a gene probe for erythropoietin messenger RNA to finally resolve the problem, but in the absence of such definitive evidence you cannot explain that in renal cell carcinoma one observes gigantic elevations of erythropoietin. Does this not leave open the possibility that tubular cells might also be a site for erythropoietin production?

Kurtz: In this study we used well-identified cells. In a carcinoma, the cell type that elaborates erythropoietin is not identified. Furthermore, it cannot be excluded that renal hypoxia is the reason for erythropoietin production in cases of a hypernephroma. There are indeed studies using tumoral cell cultures by Fisher and his group which were published last year. We cannot explain this phenomenon but we have to keep in mind that the carcinoma cells behave quite differently from normal cells and perhaps achieved characteristics which do not work in normal tubular cells.

Koch: You localized your stimulus-sensitive site within the mesangial cells themselves? Would it not be more logical to have this sensitive site within an area of the kidney which is much more sensitive to hypoxemia, like the medullary regions?

Kurtz: Yes, I agree with you but only on the first point. My opinion is that the sensing site perhaps is not the mesangium itself but the juxtaposed tubular cells, because sensing by the mesangial cell really could not explain increased erythropoietin formation by anemia. Anemia has a high arterial PO_2 and so the erythropoietin production stimulated by anemia should be measured by a structure that is able to measure the venous PO_2 and that could be tubular cells. We found when we investigated this that the tubular cells produced much more prostaglandins upon hypoxia and they produced more prostaglandin upon increased sodium chloride transport. This means that an increase in oxygen consumption of the tubular cell is capable of stimulating prostaglandin synthesis. This could be the link for both the renin release and the erythropoietin formation during hypoxia.

Ritz: Have you given thought to the possibility that the new P cells that are described by Australian authors, which very much resemble neuropeptidic cells in the juxtaglomerular apparatus, might be involved in this sensing?

Kurtz: Yes, but we have no experimental approach yet.

Koch: Thank you very much.

Contr. Nephrol., vol. 50, pp. 188–202 (Karger, Basel 1986)

Mechanisms of Abnormal Carbohydrate Metabolism in Uremia

Walter H. Hörl, Werner Riegel, Peter Schollmeyer

Department of Medicine, Division of Nephrology, University of Freiburg, Freiburg i.Br., FRG

Disturbances of carbohydrate metabolism are commonly encountered in uremic patients. Despite intensive research in this field, the origins and mechanisms of these metabolic abnormalities are still unresolved and some of the data contradictory.

Blood Glucose

Mild fasting hyperglycemia can be found in uremic patients [1] and also in acutely uremic animals [2, 3]. The evaluation of hyperglycemia is at least in part dependent on the method employed. Reaven et al. [4] determined a direct correlation between serum creatinine levels and blood glucose assayed by nonspecific methods, whereas no correlation could be made when using enzymatic methods. Similarly, Ferrannini et al. [5, 6] measured normal blood glucose values in patients on regular dialysis treatment employing specific tests, but found increased glucose values when using nonspecific methods. On the other hand, Teuscher et al. [7] achieved a direct correlation between enzymatically analyzed blood glucose and BUN in uremic patients. Hyperglycemia in uremia expressed with specific tests has been reported by several authors [2, 3, 8–10]. In contrast to these findings, spontaneous hypoglycemia in chronic uremic patients has been described probably due to substrate limitation or a deficit in renal gluconeogenesis.

Cerletty and Engbring [11] found a pathological oral glucose tolerance test (oGTT) demonstrating abnormal increases in blood glucose values of chronic uremic patients. Briggs et al. [12] detected a prolonged glucose elevation after oGTT in acute and chronic uremic patients. Sherwin et al. [13] verified hyperglycemia in patients with chronic renal failure using oGTT and demonstrated a partial normalization following dialysis. DeFronzo et al. [14] showed that initial pathological oGTT normalized 10 weeks after regular hemodialysis treatment. Lowrie et al. [15] found diminished glucose utilization in chronic uremic patients following i.v. GTT.

Insulin Action in Uremia

The early insulin response following both oral and intravenous glucose administration among uremic patients has been reported to be normal, increased or decreased. The late insulin response is uniformly elevated [16]. Lindall et al. [17] described a significant reduction in the early insulin response to glucose administration in patients with chronic renal failure treated by a Giovannetti diet compared to normal controls. In the dialyzed uremic patients without overt secondary hyperparathyroidism plasma insulin levels were not significantly different from normal controls. In contrast, when severe symptomatic hyperparathyroidism was present in dialyzed patients, plasma insulin levels were increased significantly above controls [17]. Abnormal glucose tolerance in nondialyzed patients with chronic renal failure associated with a slight decrease in initial insulin secretion after glucose administration was also observed by Horton et al. [18]. Among the possible causes of diminished early pancreatic insulin release, a depletion of potassium, lack of calcium, and elevation of magnesium have been suggested [19]. However, whether a disturbance of these electrolytes plays any major role in reduced insulin release is uncertain. Using the hyperglycemic clamp and euglycemic insulin clamp techniques, De Fronzo et al. [14] quantitated the relative contributions of impaired insulin secretion versus insulin resistance in the genesis of the glucose intolerance observed in patients with chronic renal failure [16, 20, 21]. There was no difference in the early insulin response between uremic and control subjects. However, the late plasma insulin response was significantly greater in uremic subjects versus controls. Studies carried out after 10 weeks of thrice weekly hemodialysis showed that the late plasma insulin response decreased and was no longer significantly greater than it was in controls [16]. The amount of glucose metabolized was significantly lower in the uremic patients than in normal subjects. Hemodialysis therapy resulted in a marked improvement in glucose metabolism, but did not restore glucose utilization completely to normal. The total amount of glucose metabolized (M) increased significantly postdialysis and the plasma insulin response (I) declined. Therefore the M/I ratio (a measure of tissue sensitivity to insulin) increased markedly by 80%. This was, however, still significantly less than it was in controls [16].

Sensitivity to the in vivo action of insulin was determined by the euglycemic insulin clamp technique [16, 20, 21]. Predialysis, the amount of glucose metabolized (M) by the uremic subjects and the M/I ratio were significantly less than controls. Following dialysis the amount of glucose metabolized and the M/I ratio increased significantly and was not longer different from controls. Thus, both euglycemic and hyperglycemic clamp studies indicate that tissue insensitivity to the action of insulin exists in the majority of patients with chronic renal failure and plays an important role in the glucose intolerance of uremia [16, 20, 21].

Basal hepatic glucose production and incomplete suppression of hepatic glucose production following glucose/insulin infusion was similar in uremic patients and healthy subjects [16]. Furthermore, there was no difference of nonsplanchnic glucose balance (measured by hepatic venous catheterization), hepatic glucose production (measured with tritiated glucose), and splanchnic glucose uptake in uremic patients and control subjects during the postabsorptive state and after euglycemic hyperinsulinemia [16]. The primary site of insulin resistance, however, resides in peripheral tissues. Following insulin infusion, leg glucose uptake increased in uremic patients and healthy controls. The total amount of glucose metabolized by the uremic patients was reduced by 56% compared with controls. Westervelt [22] has shown that insulin responsiveness of skeletal muscle in the uremic subject is distinctly blunted with respect to glucose and phosphorus. Data of Mondon et al. [9] also indicated that the insulin resistance of the acutely uremic rat was most likely due to decreased insulin-mediated glucose uptake by skeletal muscle, without any change in hepatic sensitivity to insulin. A decreased insulin binding on liver membranes during experimentally induced uremia has been documented by Soman and Felig [3] and Rabkin et al. [23]. No difference in insulin binding to monocytes between the uremic patients and the control subjects was observed [24]. The competition curves revealed that the half-maximal insulin binding occurred at identical insulin concentration in the two groups. The total number of insulin receptors per cell as well as the affinity of insulin for its receptor was similar in both uremic and control groups. The insulin receptor on rat hepatocytes [25] and rat soleus muscle [26] are also not reduced during chronic uremia. Therefore, the pathogenetic mechanisms responsible for the insulin resistance must be postreceptor in origin [16].

Epididymal fat has a reduced ability to transport and metabolize glucose in response to insulin, using a rat model for chronic uremia [27]. Similarly, the insulin stimulation of lipid synthesis by hepatocytes from the animals is blunted considerably [28].

The incubation of a variety of tissues with sera, obtained from chronically uremic patients, inhibits basal uptake of glucose. The insulin stimulation of glucose uptake and metabolism by rat adipocytes is also reduced following preincubation with uremic human serum [27, 29, 30]. Inhibited phosphofructokinase activity of kidney slices in the presence of uremic serum has been observed by Renner and Heintz [31]. Corresponding to these results, a decreased glucose utilization has measured in kidney slices bathed in uremic serum [32].

The glucose intolerance and impaired utilization of glucose are improved significantly after the uremic patients undergo hemodialysis. This strongly supports the hypothesis that the insulin resistance associated with uremia is due to a circulating factor(s). McCaleb et al. [33] characterized this substance as a mildly acidic, heat-stable peptide with an estimated molecular weight between 1,000 and 2,000. The resistance activity was trypsin-labile and had an apparent isoelectric point between 6 and 7. The insulin resistance activity was decreased by coincubation with the protein synthesis inhibitor, cycloheximide, indicating that this circulating small molecular weight peptide induces insulin resistance in uremia by a protein synthesis dependent mechanism [33].

The hepatic insulin degradation and clearance rate during uremia is unchanged [3, 4, 9, 23]. According to studies by Ferrannini et al. [5, 6] and Navalesi et al. [34], the degradation of total body insulin is decreased under uremic conditions. Quite possibly, the loss of renal parenchyma plays an important role in the reduced insulin breakdown of the uremic patients [35]. On the other hand, Mondon et al. [9] and Rabkin et al. [23] were able to demonstrate that especially muscle insulin degradation during uremia is clearly reduced.

Elevated immunoreactive insulin plasma levels have been observed in fasting uremic patients [10, 13, 14] and in experimentally induced uremia [3, 4, 8, 36].

Glucagon Action in Uremia

Hyperglucagonemia in experimentally induced uremia [3, 8, 35, 37] and in patients with renal insufficiency [13, 38] has been well documented. This hyperglucagonemia remains unaffected by glucose infusion in chronic uremic patients. A direct correlation exists between plasma glucagon and creatinine levels. Sherwin et al. [13] reported that there are no differences between the raised glucagon values of dialyzed and nondialyzed chronic uremic patients. However, dialysis normalized the glycemic response to glucagon. Neither the oGTT nor alanine administration may influence hyperglucagonemia [13]. Emmanouel et al. [37] determined hyperglucagonemia in rats following bilateral nephrectomy or ureter ligation, but not in uremia caused by autoinfusion of urine. Consequently, these authors believe that hyperglucagonemia is not a result of uremia per se. The elevated plasma glucagon concentration results from decreased clearance by the kidney [13, 37, 39] and consists of increased levels of proglucagon as well as the biologically active 3,500 molecular weight species. Hyperglucagonemia in renal insufficiency is normalized after kidney transplantation [40].

The results of Soman and Felig [3] suggest a hypersensitivity to glucagon in uremia. These authors determined an increased binding to glucagon on liver membranes parallel with an augmented adenylate cyclase activity. Sherwin et al. [13] described an exaggerated glycemic response to infused glucagon in chronic uremic patients. In contrast, Mondon et al. [41] presented evidence for glucagon resistance in the liver during experimental acute and chronic uremia. Perfusion of livers from control rats with glucagon-containing perfusate leads to a significant and sustained elevation of perfusate glucose concentration as compared to the saline perfusion. In contrast, glucose concentration did not increase when livers from acutely uremic rats were perfused with glucagon [41]. However, perfused livers from acutely uremic rats responded to dibutyryl-cAMP, despite being resistant to glucagon [41]. Glucagon, which did not increase hepatic glucose efflux, was also incapable of increasing urea nitrogen production in perfused livers of acutely uremic rats. In contrast, urea formation increased and glycogen content decreased when livers from acutely uremic rats were perfused with dibutyryl-cAMP [41]. Glucagon infused at rates of 6 ng/min/kg rat resulted in minimal and equivalent increases in hepatic glucose outflow and cAMP accumulation when livers from acutely uremic and control rats were perfused. However, at glucagon infusion rates of 18 ng/min/kg, glucose efflux from perfused livers of acutely and chronically uremic rats was significantly reduced and cAMP accumulation was also significantly lower [42]. Thus, glucose intolerance in uremic rats does not appear to be due to increased hepatic glucose output resulting from increased sensitivity to glucagon [42].

Gluconeogenesis

Glucagon [43] and catecholamines [44] are known to stimulate hepatic gluconeogenesis by a various spectrum of mechanisms, e.g. hepatic uptake of precursors, formation of cAMP and stimulation of key enzymes of gluconeogenesis. Lacy [45] showed in liver perfusion experiments of acutely uremic rats that the uptake of different amino acids is stimulated, whereas branched chain amino acids were released. Fröhlich et al. [46] found that the uptake of amino acids but not of lactate and pyruvate is increased in acutely uremic rats. Even the uptake of the nonmetabolizable amino acid α-aminoisobutyrate was enhanced. It was shown that serine plays a special role as a substrate for the additional glucose formation in acutely uremic rats, probably mediated by an activation of serine dehydratase. An increased supply of amino acids within the liver cell is to be expected both from stimulation of endogenous proteolysis and from accelerated amino acid transport [47]. This assumption is supported by studies of Kahlhan et al. [48] showing increased glucose carbon recycling in chronically uremic patients, while glucose oxidation is decreased.

The activation of liver gluconeogenesis in acute renal failure is confirmed by studies in isolated rat hepatocytes [49–52]. However, this effect is not uniform and depends upon the methods used for induction of acute uremia as well as on the substrate used in the incubation medium. Glucose production from serine by isolated hepatocytes of binephrectomized and ureter-ligated rats and animals with uranyl nitrate-induced acute renal failure was markedly enhanced compared with control or sham-operated animals. In contrast, a less but significant increase of glucose synthesis from pyruvate or dihydroxyacetone was observed after binephrectomy or ureter ligation. Inhibition of glucose production was found in hepatocytes of rats with uranyl nitrate-induced acute renal failure in the presence of dihydroxyacetone [49, 50].

Incubation with glucagon, adrenaline or dibutyryl-cAMP resulted in high activation of glucose synthesis in sham-operated, binephrectomized

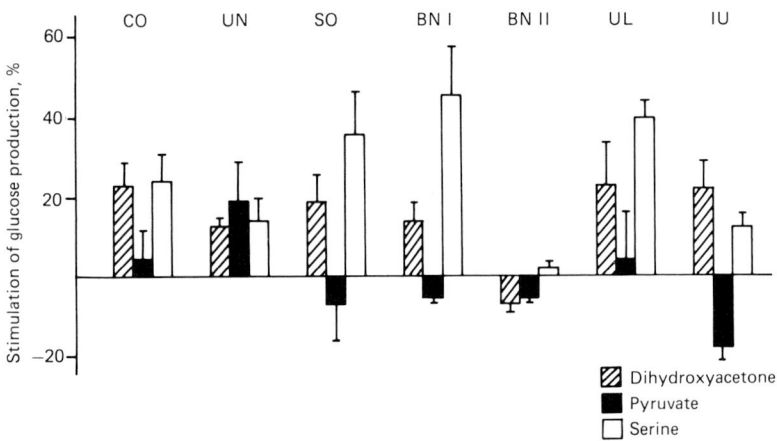

Fig. 1. Effect of glucagon (28 nmol/l) on glucose production in isolated liver cells of different rat models of acute uremia. UN = Uranyl nitrate-treated rats; BN = bilaterally nephrectomized rats; UL = ureter-ligated rats; IU = rats made ischemic; CO = untreated controls; SO = sham operation. Mean values ± SEM from 8 experiments.

(50%) und ureter-ligated rats using serine or dihydroxyacetone as substrates [51]. In contrast, in half of the experiments with hepatocytes prepared from bilaterally nephrectomized rats (BN II) we could not detect any effects of the hormones used. However, basal glucose production and hepatocyte viability were identical in both groups of binephrectomized animals (BN I and BN II) in the presence of the different substrates used.

At present it remains unclear why responses to glucagon were obtained only in BN I (fig. 1). Therefore, it cannot be generally concluded that glucagon resistance may exist in the liver of acutely uremic rats [41].

Moreover, further parameters were considered which may play a key role in the regulation of liver cell metabolism: (a) hormonal condition of the animals; (b) function of hormonal receptors; (c) activity of key enzymes of gluconeogenesis; (d) suitable carbon substrate supply; (e) reducing equivalents, and (f) high-energy phosphate content.

A significant increase of ATP content and energy charge was observed in hepatocytes from ureter-ligated rats in the presence of serine, pyruvate and dihydroxyacetone, compared with binephrectomized animals, whereas the gluconeogenetic rate from these substrates was nearly identical. Lactate/pyruvate ratio characterizing cytosolic reducing equivalents in hepatocytes of bilaterally nephrectomized and especially of ureter-ligated rats in-

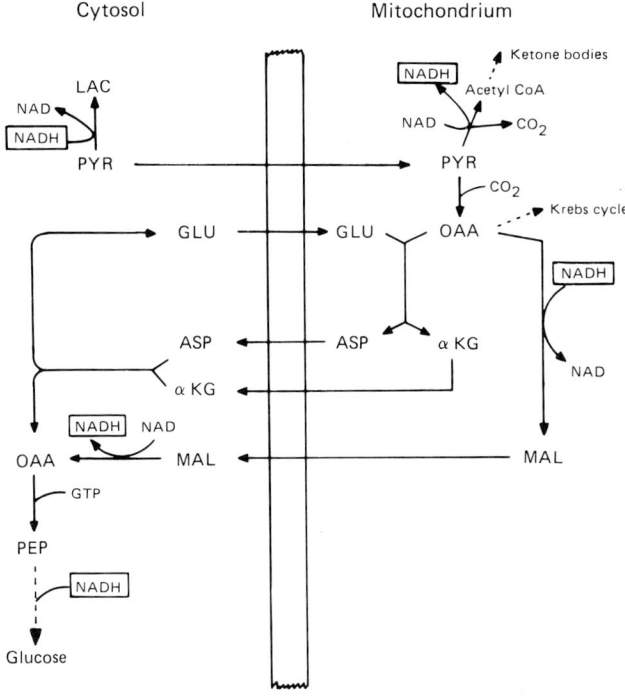

Mitochondrial – cytosolic transport system of reducing equivalents
during gluconeogenesis from pyruvate

Fig. 2. A model for the relationship between cytosolic and mitochondrial reducing
equivalents. αKG = α-ketoglutarat; ASP = aspartate; GLU = glutamate; LAC = lactate;
MAL = malate; PEP = phosphoenolpyruvate; PYR = pyruvate; OAA = oxalacetoace-
tate.

dicates lack of reducing equivalents in the cytosol of acutely uremic animals.
β-Hydroxybutyrate/acetoacetate ratio characterizing the mitochondrial
redox state was significantly lower in hepatocytes of acutely uremic rats.
This quotient decreased during incubation indicating accumulation of reducing equivalents into the mitochondria [53]. Stimulated ketogenesis in
hepatocytes of acutely uremic rats in the presence of serine, pyruvate and
dihydroxyacetone provides mitochondrial NADH and can be considered as
one reason for this mechanism. Therefore, we conclude that hepatic
gluconeogenesis of acutely uremic rats are limited by a lack of cytosolic
reducing equivalents independently of cell energy supply [53]. The rate-
limiting role of cytosolic reducing equivalents is summarized in figure 2.

Feeding with a mixed, low-protein diet cannot prevent the rapid decline in liver glycogen in acutely uremic rats. In binephrectomized rats fed with lactate combined with a low-protein diet also a drastically reduced liver glycogen content was observed. However, an approximately 30-fold increase in liver glycogen was measured with lactate succeeding sham operation after 24-hour fasting. Comparable high liver glycogen was determined after serine administration in acutely uremic and sham-operated rats [49]. One possible explanation for serine utilization in acute uremia is the fact that this amino acid in contrast to lactate and other substrates can be metabolized independently of the redox state of liver cells.

Glycogen Metabolism

Bergström and Hultman [54] found a distinct fall in muscle glycogen of patients with acute renal failure. Although, this decrease was not dependent on the cause of acute renal failure, the most impressive results were obtained by postoperative acute renal failure. To elucidate the pathogenesis of this phenomenon, we investigated glycogen metabolizing enzymes in acutely uremic rats.

Muscle phosphorylase *a*, the active form of the glycogen-degrading enzyme, was maximally activated in both binephrectomized as well as in sham-operated fasting rats. Sixty hours after operation, the values of phosphorylase *a* activity were in the range seen before induction of acute uremia [55].

Twenty hours after bilateral nephrectomy or sham operation, only 10% of glycogen synthetase activity was found. Forty and sixty hours after operation only a slight increase of glycogen synthetase activity was observed [55]. These data indicate activation of glycogen breakdown and inhibition of glycogen synthesis in skeletal muscle of acutely uremic rats. Nevertheless, activation of glycogenolysis observed during acute uremia must be completely different compared to sham operation.

Whilst it was possible to isolate an intact protein-glycogen complex 60 h after sham operation in animals on a high-calorie, low-protein diet, keto acid supplements and propranolol, this could not be done in acutely uremic rats under the same conditions [56]. This study clearly demonstrates that glycogenolysis caused by acute uremia is specifically different from a purely stress-induced stimulation of glycogen breakdown. There are two possible explanations for these results: (1) in acute uremia glycogenolysis in skeletal muscle is not activated via the β-adrenergic system, or (2) the β-receptor may be altered probably by proteolysis and therefore is not blocked by propranolol. There is some evidence suggesting an important role of proteases in acute renal failure [57−60].

The favorable effects of serine on muscle glycogen metabolism that we observed during acute uremia are astounishing. An indirect anabolic influence on carbohydrate metabolism of skeletal muscle may be induced by the stimulation of liver gluconeogenesis by serine [52].

Isolated hindquarters of bilaterally nephrectomized and sham-operated rats were perfused in the presence and absence of ^{14}C-labelled serine, respectively. After a perfusion period of 30 min ^{14}C-serine was 4,074 ± 270 dpm/ml in the perfusion medium of sham-operated animals and decreased to 2,800 ± 190 dpm/ml in the medium of acutely uremic rats. Muscle glycogen concentration in sham-operated animals was 1.10 ± 0.04 mg/g wet weight in the absence and 1.03 ± 0.11 mg/g in the presence of serine. In contrast, in acutely uremic rats there was a glycogen concentration of 0.57 ± 0.09 mg/g in the absence of serine. Glycogen was increased in the presence of serine in the perfusion medium, the value being 1.50 ± 0.13 mg glycogen/g wet weight. Incorporation of labelled serine into skeletal muscle glycogen was significantly higher in acutely uremic animals (15 ± 0.5 μmol/g glycogen) than in sham-operated animals (10 ± 0.4 μmol/g). The results are compatible with the hypothesis that serine increases muscle glycogen synthesis in acute uremia [61].

In comparison to sham-operated controls Penpargkul and Scheuer [62] observed increased heart glycogen levels in fasting and fed rats 24 h after ureter ligation. Interestingly, fasting uremic rats displayed higher heart glycogen concentrations than fed uremic animals. Epinephrine stimulated glycogenolysis in sham-operated as well as in acutely uremic animals, whereas no differences in lactate production could be observed. Since the activities of phosphorylase and glycogen synthetase in heart muscle were lowered, these authors assumed a reduced myocardial glycogen turnover during acute uremia.

Our experiments on fasting binephrectomized and ureter-ligated rats showed an increase of myocardial glycogen as well as total carbohydrate content 24 h after induction of acute uremia in comparison to untreated and sham-operated animals [49].

Forty-eight hours after operation there was a decrease of heart muscle glycogen, whereas the myocardial glucose concentration in acutely uremic rats was enhanced. The fact that glycogen and/or glucose concentrations of the uremic heart muscle revealed a further rise during starvation suggest that these substrates may play a protective role in the heart metabolism.

In ureter-ligated rats, Reaven and Reaven [63] demonstrated the key role of the kidney in the development of uremic hyperglycemia. According to these investigations metabolic acidosis facilitates the increased gluconeogenetic activity of the kidney. The loss of renal parenchyma causes a converse effect, namely hypoglycemia.

In order to evaluate the potential role of hyperkalemia und metabolic acidosis on the disturbances of carbohydrate metabolism normally seen in uremia, a specific model of acute uremia devoid of hyperkalemia and severe metabolic acidosis was chosen. Therefore, rats were deprived of potassium prior to induction of acute uremia. Potassium depletion caused a significant decrease of muscle and liver glycogen due to activation of phosphorylase kinase, whereas glycogen concentration in heart muscle was unchanged and elevated in the kidney of sham-operated and ureter-ligated animals. In contrast, glucose concentrations were enhanced in the liver and the kidney, unchanged in heart muscle and decreased in skeletal muscle. We conclude that carbohydrate metabolism of heart muscle is not affected by changes of plasma potassium and acid-base status. Hypokalemia and metabolic alkalosis prior to induction of acute uremia, however, cause further stimulation of glycogenolysis of skeletal muscle and the liver [64].

References

1   DeFronzo, R.A.; Andres, R.; Edgar, P.; Walter, W.G.: Carbohydrate metabolism in uremia: a review. Medicine 52: 469–481 (1973).
2   Reaven, G.M.; Reaven, P.D.: Development of fasting hyperglycemia in uremic rats. Metabolism 26: 1251–1256 (1977).
3   Soman, V.; Felig, P.: Glucagon and insulin binding to liver membranes in a partially nephrectomized uremic rat model. J. clin. Invest. 60: 224–232 (1977).
4   Reaven, G.M.; Weisinger, J.R.; Swenson, R.S.: Insulin and glucose metabolism in renal insufficiency. Kidney int. 6: suppl. 1, pp. S63–S69 (1974).
5   Ferrannini, E.; Pilo, A.; Tuoni, M.: The response to intravenous glucose of patients on maintenance hemodialysis. Effect of dialysis. Metabolism 28: 125–136 (1979).
6   Ferrannini, E.; Pilo, A.; Navalesi, R.; Citti, L.: Insulin kinetics and glucose-induced insulin delivery in chronically dialyzed subjects: acute effects of dialysis. J. clin. Endocr. Metab. 49: 15–22 (1979).
7   Teuscher, A.; Fankhauser, S.; Küffer, F.R.: Untersuchungen zum Kohlenhydratstoffwechsel bei Niereninsuffizienz. Klin. Wschr. 41: 706–715 (1963).
8   Bilbrey, G.L.; Faloona, G.R.; White, M.G.; Knochel, J.P.: Hyperglucagonemia of renal failure. J. clin. Invest. 53: 841–847 (1974).

9   Mondon, C.E.; Dolkas, C.B.; Reaven, G.M.: The site of insulin resistance in acute
    uremia. Diabetes 27: 571–576 (1978).
10 Rubenfeld, S.; Garber, A.J.: Abnormal carbohydrate metabolism in chronic renal fail-
    ure. J. clin. Invest. 62: 20–28 (1978).
11 Cerletty, J.M.; Engbring, N.H.: Azotemia and glucose intolerance. Ann. intern. Med.
    66: 1097–1108 (1967).
12 Briggs, J.D.; Buchanan, K.D.; Luke, R.G.; McKiddie, M.T.: Role of insulin in glucose
    intolerance in uraemia. Lancet i: 462–464 (1967).
13 Sherwin, R.S.; Bastl, C.; Finkelstein, F.O.; Fisher, M.; Black, H.; Hendler, R.; Felig,
    P.: Influence of uremia and hemodialysis on the turnover and metabolism of glucagon.
    J. clin. Invest. 57: 722–731 (1976).
14 DeFronzo, R.A.; Jordan, D.T.; Rowe, J.W.; Andres, R.: Glucose intolerance in
    uremia. Quantification of pancreatic beta cells sensitivity to glucose and tissue sensitivity
    to insulin. J. clin. Invest. 62: 425–435 (1978).
15 Lowrie, E.G.; Soeldner, J.S.; Hampers, C.L.; Merril, J.P.: Glucose metabolism and in-
    sulin secretion in uremic, prediabetic, and normal subjects. J. Lab. clin. Med. 76:
    603–615 (1970).
16 DeFronzo, R.A.; Smith, D.; Alvestrand, A.: Insulin action in uremia. Kidney int. 24:
    suppl. 16, pp. S102–S114 (1983).
17 Lindall, A.; Carmena, R.; Cohen, S.; Comty, C.: Insulin hypersecretion in patients on
    chronic hemodialysis. Role of parathyroids. J. clin. Endocr. 32: 653–658 (1971).
18 Horton, E.S.; Johnson, C.; Lebovitz, H.E.: Carbohydrate metabolism in uremia. Ann.
    intern. Med. 68: 63–74 (1968).
19 Fröhlich, J.; Schollmeyer, P.; Gerok, W.: Carbohydrate metabolism in renal failure.
    Am. J. clin. Nutr. 31: 1541–1546 (1978).
20 DeFronzo, R.A.: Pathogenesis of glucose intolerance in uremia. Metabolism 27: suppl.
    2, pp. 1866–1880 (1978).
21 DeFronzo, R.A.; Alvestrand, A.: Glucose intolerance in uremia: site and mechanism.
    Am. J. clin. Nutr. 33: 1438–1445 (1980).
22 Westervelt, F.B.: Insulin effect in uremia. J. Lab. clin. Med. 74: 79–84 (1969).
23 Rabkin, R.; Unterhalter, S.; Duckworth, W.C.: Effect of prolonged uremia on insulin
    metabolism by isolated liver and muscle. Kidney int. 16: 433–439 (1979).
24 Smith, D.; DeFronzo, R.A.: Insulin resistance in uremia mediated by post binding de-
    fects. Kidney int. 22: 54–62 (1982).
25 Kaufmann, J.M.; Caro, J.F.: Insulin resistance in uremia. J. clin. Invest. 71: 698–708
    (1983).
26 Smith, D.; DeFronzo, R.A.: Insulin, glucagon and thyroid hormone. Renal Endocr. 11:
    367–428 (1983).
27 Maloff, B.L.; McCaleb, M.L.; Lockwood, D.H.: Cellular basis of insulin resistance in
    chronic uremia. Am. J. Physiol. 245: E178–E184 (1983).
28 Caro, J.F.; Lanza-Jacoby, S.: Insulin resistance in uremia. J. clin. Invest. 72: 882–892
    (1983).
29 Dzurik, R.: Metabolic alterations caused by uremia. Proc. Eur. Dial. Transplant Ass. 17:
    577–586 (1980).
30 McCaleb, M.L.; Mevorach, R.; Freeman, R.B.; Izzo, M.S.; Lockwood, D.H.: Induc-
    tion of insulin resistance in normal adipose tissue by uremic human serum. Kidney int.
    25: 416–421 (1984).

31 Renner, D.; Heintz, R.: Der Einfluss urämischen Serums auf die Glucoseneubildung und auf die aktivierte Essigsäure in Gewebeschnitten. Untersuchungen zur urämischen Intoxikation. Klin. Wschr. *51:* 82–87 (1973).

32 Dzurik, R,; Hupkova, V.; Cernacek, P.; Valovicova, E.: The isolation of an inhibitor of glucose utilization from the serum of uraemic subjects. Clinica chim. Acta *46:* 77–83 (1973).

33 McCaleb, M.L.; Izzo, M.S.; Lockwood, D.H.: Characterization and partial purification of a factor from uremic human serum that induces insulin resistance. J. clin. Invest. *75:* 391–396 (1985).

34 Navalesi, R.; Pilo, A.; Lenzi, S.; Donato, L.: Insulin metabolism in chronic uremia and in the anephric state: effect of dialytic treatment. J. clin. Endocr. Metab. *40:* 70–85 (1975).

35 Katz, A.I.; Emmanouel, D.S.: Metabolism of polypeptide hormones by the normal kidney and in uremia. Nephron *22:* 69–80 (1978).

36 Arnold, W.C.; Holliday, M.A.: Tissue resistance to insulin of amino acid uptake in acutely uremic rats. Kidney int. *16:* 124–129 (1979).

37 Emmanouel, D.S.; Jaspan, J.B.; Kuku, S.F.; Rubenstein, A.H.; Katz, A.I.: Pathogenesis and characterization of hyperglucagonemia in the uremic rat. J. clin. Invest. *58:* 1266–1272 (1976).

38 Kuku, S.F.; Jaspan, J.B.; Zeidler, A.; Katz, A.I.; Rubenstein, A.H.: Heterogenity of plasma glucagon. Circulating components in normal subjects and patients with chronic renal failure. J. clin. Invest. *58:* 742–750 (1976).

39 Bastl, C.; Finkelstein, F.O.; Sherwin, R.; Hendler, R.; Felig, P.; Hayslett, J.P.: Renal extraction of glucagon in rats with normal and reduced kidney function. Am. J. Physiol. *233:* F67–F71 (1977).

40 Bilbrey, G.L.; Faloona, G.R.; White, M.G.; Atkins, C.; Hull, A.R.; Knochel, J.P.: Hyperglucagonemia in uremia. Reversal by renal transplantation. Ann. intern. Med. *82:* 525–528 (1975).

41 Mondon, C.E.; Reaven, G.M.: Evaluation of enhanced glucagon sensitivity as the cause of glucose intolerance in acutely uremic rats. Am. J. clin. Nutr. *33:* 1456–1460 (1980).

42 Mondon, C.E.; Marcus, R.; Reaven, G.M.: Role of glucagon as a contributor to glucose intolerance in acute and chronic uremia. Metabolism *31:* 374–379 (1982).

43 Exton, J.; Park, C.: D. The stimulation of gluconeogenesis from lactate by epinephrine, glucagon, and cyclic 3',5'-adenylate in the perfused rat liver. Pharmac. Rev. *18:* 181–188 (1966).

44 Hue, L.; Felin, J.; Hers, H.: Control of gluconeogenesis and of enzymes of glycogen metabolism in isolated rat hepatocytes. A parallel study of phenylephrine and of glucagon. Biochem. J. *176:* 791–797 (1978).

45 Lacy, W.: Uptake of individual amino acids by perfused rat liver. Effect of acute uremia. Am. J. Physiol. *219:* 649–653 (1970).

46 Fröhlich, J.; Hoppe-Seyler, G.; Schollmeyer, P.; Maier, K.; Gerok, W.: Possible sites of interaction of acute renal failure with amino acid utilization for gluconeogenesis in isolated perfused rat liver. Eur. J. clin. Invest. *7:* 261–268 (1977).

47 Fröhlich, J.; Schölmerich, J.; Hoppe-Seyler, G.; Maier, K.; Talke, H.; Schollmeyer, P.; Gerok, W.: The effect of acute uraemia on gluconeogenesis in isolated perfused rat livers. Eur. J. clin. Invest. *4:* 453–458 (1974).

48 Kalkhan, S.; Ricanati, E.; Tsemg, K.; Savin, S.: Glucose turnover in chronic uremia: in-

creased recycling with diminished oxidation of glucose. Metabolism *32:* 1155−1162 (1983).

49 Hörl, W.H.; Stepinski, J.; Heidland, A.: Carbohydrate metabolism and uraemia − mechanisms for glycogenolysis and gluconeogenesis. Klin. Wschr. *58:* 1051−1064 (1980).

50 Stepinski, J.; Hörl, W.H.; Heidland, A.: The gluconeogenetic ability of hepatocytes in various types of acute uraemia. Nephron *31:* 75−81 (1982).

51 Riegel, W.; Stepinski, J.; Hörl, W.H.; Heidland, A.: Effect of hormones on hepatocyte gluconeogenesis in different models of acute uraemia. Nephron *32:* 67−72 (1982).

52 Riegel, W.; Stepinski, J.; Münchmeyer, M.; Heidland, A.; Hörl, W.H.: Effect of serine on gluconeogenic ability of hepatocytes in acute uremia. Kidney int. *24:* suppl. 16, pp. S48−S51 (1983).

53 Riegel, W.; Hörl, W.H.: Role of energy charge and redox state for hepatocyte gluconeogenesis of acutely uremic rats. Nephron *40:* 206−212 (1985).

54 Bergström, J.; Hultman, E.: Glycogen content of skeletal muscle in patients with renal failure. Acta med. scand. *186:* 177−181 (1969).

55 Hörl, W.H.; Heidland, A.: Glycogen metabolism in muscle in uremia. Am. J. clin. Nutr. *33:* 1461−1467 (1980).

56 Schäfer, R.M.; Hörl, W.H.; Heidland, A.: Isolation of protein-glycogen complexes from rat skeletal muscle in acute uremia. Role of serine. Proc. Soc. exp. Biol. Med. *169:* 519−526 (1982).

57 Hörl, W.H.; Heidland, A.: Enhanced proteolytic activity − cause of protein catabolism in acute renal failure. Am. J. clin. Nutr. *33:* 1423−1427 (1980).

58 Hörl, W.H.; Stepinski, J.; Gantert, C.; Hörl, M.; Heidland, A.: Evidence for the participation of proteases on protein catabolism during hypercatabolic renal failure. Klin. Wschr. *59:* 751−759 (1981).

59 Hörl, W.H.; Gantert, C.; Auer, I.O.; Heidland, A.: In vitro inhibition of protein catabolism by alpha$_2$-macroglobulin in plasma from a patient with posttraumatic acute renal failure. Am. J. Nephrol. *2:* 33−37 (1982).

60 Hörl, W.H.; Stepinski, J.; Schäfer, R.M.; Wanner, C.; Heidland, A.: Role of proteases in hypercatabolic patients with renal failure. Kidney int. *24:* suppl. 16, pp. S37−S42 (1983).

61 Djovkar, A.; Hörl, W.H.; Heidland, A.: Glycogen synthesis from serine in isolated perfused hindquarters of acutely uremic rats. Nephron *34:* 164−167 (1983).

62 Penpargkul, S.; Scheuer, J.: Regulation of glycogen metabolism in acute uremic hearts. Metabolism *23:* 631−644 (1974).

63 Reaven, G.M.; Reaven, P.D.: Development of fasting hyperglycemia in uremic rats. Metabolism *26:* 1251−1256 (1977).

64 Hörl, W.H.; Schaefer, R.M.; Heidland, A.: Acute uremia following dietary potassium depletion,. II. Effect on tissue carbohydrate composition. Mineral Electrolyte Metab. (submitted for publication).

Prof. Dr. Dr. Walter H. Hörl, Medizinische Universitätsklinik, Hugstetter Strasse 55, D−7800 Freiburg i.Br. (FRG)

Discussion

Kokot: Thank you very much, Prof. Hörl, for your excellent presentation. It is now open for discussion. Are there any comments, remarks?

Wizemann: You mentioned the possible role of insulinotropic substances from the gut. Is there any?

Hörl: In my topic I discussed the possible role of insulin, glucagon, adrenaline or PTH.

Wizemann: No, you did not mention it but there are some insulinotropic hormones coming from the gut, like GIP and others. Is there any role of those hormones in uremia?

Hörl: As I know from the literature there are no data concerning the potential metabolic effects of these hormones on the alterations of carbohydrate in uremia.

Kokot: Is glucagon-induced gluconeogenesis dependent upon calcium concentration in the medium both in the uremic and in normal animals?

Hörl: In our experiments we kept the calcium concentration constant. In experiments with phosphorous-depleted animals which we have done together with Prof. Ritz, the effect of glucagon on hepatic gluconeogenesis was reduced.

Kokot: Any more questions? If not, thank you very much once more.

Contr. Nephrol., vol. 50, pp. 203–210 (Karger, Basel 1986)

Partial Correction of Lipid Disturbances by Insulin in Experimental Renal Failure

J.B. Roullet, B. Lacour, T. Drueke[1]

Département de Néphrologie et Inserm U-90, Hôpital Necker, Paris, France

In chronic renal failure the metabolism of lipoproteins and carbohydrates is disturbed including impaired glucose tolerance, peripheral insulin resistance and accumulation of triglyceride(TG)-rich lipoproteins [1, 7, 9, 23, 25]. A deficiency of lipoprotein lipase (LPL) has been reported to play an important role in the accumulation of TG-rich lipoproteins [5, 11, 14, 24]. The latter are abnormally enriched in cholesterol [23] and may be of pathogenetic importance in the accelerated atherosclerosis of chronically uremic patients. Thus, cholesterol-enriched very low density lipoproteins (β-VLDL), induced by cholesterol feeding to experimental animals, resulted in atherosclerosis [18].

Since insulin is essential for LPL synthesis [6], we tested the hypothesis whether chronic insulin infusion into uremic rats could lead to a correction of their lipid disturbances and thus represent a potential means of reducing 'atherogenic' VLDLs.

Experimental Model

Three groups of rats were studied: one group of 21 control rats (C) and two groups of uremic rats: 19 uremic rats (U) without hormone supplementation and 23 uremic rats that received chronic insulin supplementation (U+). Chronic renal insufficiency was created by cortical electrocoagulation and subsequent contralateral nephrectomy [3]. Pig insulin (Actrapid®, Novo Laboratories) was administered by means of an osmotically driven minipump (0.46 U insulin per day) continuously over the 35-day experimental period. All rats were fed standard laboratory chow ad libitum.

[1] The authors wish to thank Ms. F. Buhot for valuable secretarial help.

On day 15, animals underwent ether anesthesia in the nonfasted state and 400 μl of blood were collected in the jugular vein in order to determine serum urea, creatinine and glucose. On day 30, hepatic triglyceride secretion rate (TGSR) was determined using Triton WR 1339 as described by Bagdade et al. [2]. On day 35, all rats were killed by decapitation after a 16-hour fast.

VLDL (density < 1.006) were isolated according to Havel et al. [12] and their protein content determined by the method of Lowry et al. [17]. Adipose tissue LPL was measured as described by Nilsson-Ehle and Schotz [22]. Serum insulin was quantitated with an INSIK-3 kit (Sorin Laboratories, Biomedica S.p.a., Italy). Serum β-hydroxybutyrate (BOB) and nonesterified fatty acids (NEFA) were quantitated as previously described [16, 17]. All other parameters were determined with standard biochemistry methods. Results were expressed in terms of mean values ± SEM.

Results

The mean body weight gain of U and U+ rats (55 and 46 g, respectively) was less than that of C rats (85 g) over the 5-week experimental period. Serum creatinine concentrations of U and U+ rats (104 and 110 μM) was significantly increased when compared to that of C animals (68 μM). Serum glucose was determined twice over the experimental period. On day 15, rats being under ether anesthesia and without any fast, no difference was observed between the three groups. On day 35, after a 16-hour fast and without anesthesia, glucose levels of U and U+ rats were significantly decreased (5.4 and 3.7 mM) when compared to C rats (6.3 mM), the mean concentrations of U+ rats being significantly lower than that of U rats. At the end of the experimental period, U rats exhibited a significantly reduced serum insulin concentration (17.0 ± 0.6 mU/l) in the fasted state while serum insulin of U+ rats (23.4 ± 1.7) was similar to that of C rats (20.4 ± 1.2). No difference was observed between the serum lactate concentration of the three groups.

Table I shows serum total lipids on day 35, i.e. day of sacrifice. It indicates that serum TG levels of U+ rats were identical to those of C rats whereas TG concentrations of U rats were significantly increased. For total cholesterol and phospholipid levels, no difference was observed between U+ and U rats, both having an increase in their serum cholesterol and phospholipid concentrations when compared to C rats.

Table II shows that on day 35, serum VLDL-TG, cholesterol, phospholipid and protein concentrations of U rats were significantly increased when compared to those of C and U+ rats. C and U+ rats were similar for these parameters. The ratio of cholesterol to TG contained in VLDL was increased in U rats and near normal in U+ rats.

Table I. Serum total TG, cholesterol and phospholipids concentrations in control (C), uremic (U) and uremic with insulin (U+) rats after a 16-hour fast at the time of sacrifice (mean ± SEM)

	C (n = 21)	U (n = 19)	U+ (n = 23)
TG, g/l	0.69 ± 0.03*	1.04 ± 0.07*	0.69 ± 0.07
Cholesterol, g/l	0.53 ± 0.02*	0.91 ± 0.04	0.95 ± 0.05*
Phospholipids, g/l	0.83 ± 0.03*	1.57 ± 0.04	1.59 ± 0.06*

* $p < 0.05$. C rats were compared with U rats, U rats with U+ rats and U+ rats with C rats.

Table II. VLDL-TG, VLDL-cholesterol, and VLDL-phospholipids and VLDL-protein concentrations in control (C), uremic (U) and uremic with insulin (U+) rats after a 16-hour fast at the time of sacrifice (mean ± SEM)

	C (n = 20)	U (n = 19)	U+ (n = 19)
VLDL-TG, mg/l	445 ± 38*	804 ± 65*	410 ± 36
VLDL-cholesterol, mg/l	13 ± 3*	43 ± 8*	16 ± 3
VLDL-phospholipids, mg/l	70 ± 9*	142 ± 5*	83 ± 8
VLDL-protein, mg/l	272 ± 16*	374 ± 23*	290 ± 23
VLDL-cholesterol/ VLDL-TG ratio	0.031 ± 0.006*	0.054 ± 0.006*	0.037 ± 0.005

* $p < 0.05$. C rats were compared with U rats, U rats with U+ rats and U+ rats with C rats.

Epididymal fat pad weight of U rats (3.1 g ± 0.2 g) was significantly reduced when compared to that of C rats (6.4 ± 0.4 g). Chronic insulin infusion did not restore the epididymal fat weight of U+ rats to normal (4.0 ± 0.3 g) but increased it significantly by comparison with U rats. In addition, U+ rats had lower concentrations of serum NEFA, glycerol and BOB when compared to that of U and C rats. The TGSR of U+ rats (0.44 ± 0.03 μmol/min) was significantly lower than that of U and C rats (0.58 ± 0.03 and 0.57 ± 0.05 μmol/min, respectively). Moreover, the LPL activity of U+ rats (860 ± 150 mU per total epididymal fat mass) was found to be similar to that of C rats (705 ± 85) whereas the LPL activity of U rats (460 ± 60) was decreased by comparison to that of C and U+ rats.

Discussion

The accumulation of TG-rich lipoproteins in uremic rats [2, 13, 26] is attributed mainly to a decreased removal from plasma, related to deficient LPL activity and not to an increased hepatic synthesis [2]. This finding is confirmed in the present study as LPL activity of adipose tissue was low and TGSR normal in the group of untreated uremic rats.

The disturbed lipoprotein metabolism could, at least in part, be due to the abnormal carbohydrate metabolism of uremia since insulin is the principal hormone that stimulates LPL synthesis [27], and since a peripheral and hepatic resistance to insulin action has been demonstrated in chronically uremic rats [15, 19]. Our experimental model of chronic renal failure showed a decrease in circulating insulin associated with normoglycemia in the anesthetized fed state and moderate hypoglycemia after a 16-hour fast. Thus, our model appears to represent an insulin-deficient state. This situation is consistent with the observations of Bagdade et al. [2] and Heuck et al. [13] but in contrast with others [15, 19]. It cannot be a consequence of the chronic food deprivation of uremia since we have shown [26] that sham-operated pair-fed rats exhibited normal insulin levels and normal adipose LPL activity, but could be related to some alteration of the insulin secretion by pancreatic β-cells, as also observed in patients with chronic renal failure [8].

In the present study, the long-term infusion of insulin to uremic rats led to a correction of the activity of LPL and thus a normalization of the clearance and composition of TG-rich lipoproteins. We chose to administer a moderate dose of insulin sufficient to provide normal levels of the circulating hormone but low enough not to induce hyperinsulinemia. The infused dose proved to be adequate since no hypoglycemia occurred in the fed state. As a consequence of a possible effect of insulin on protein intake, uremic with insulin rats exhibit higher serum urea concentration when compared to animals of comparable degree of uremia. In contrast to what has been reported with higher insulin doses [21], no increase in body weight gain was observed. The only significant morphological change was an increase in epididymal fat pad weight. This datum, together with the normalization of LPL activity, can be explained by the lipogenic action of insulin, and in particular its stimulating effect on LPL synthesis. These results are consistent with the hypothesis of an insulin-related decrease of LPL in chronic uremia.

As to TG and VLDL, chronic insulin infusion led to a normalization of their respective serum concentrations. This effect appears to be the result of a combined action of insulin: (1) the hormone restores the catabolism of VLDL by improving LPL activity, thus leading to circulating VLDL with normal cholesterol and TG contents and (2) it decreases the hepatic synthesis of TG (significant decrease of TGSR). The latter result is surprising if one considers that insulin stimulates TG synthesis via hepatic carnitine acyltransferase in normal rats [4]. In fact, we observed that chronic insulin infusion decreased plasma concentration of NEFA (and glycerol) which are the main products of peripheral lipolysis. Because of its lowering effect on peripheral NEFA flux and because NEFA regulate mainly liver TG synthesis, the global effect of chronic insulin infusion in uremic rats is a decreased production of TG by the liver.

Conclusion

The long-term infusion of insulin into uremic rats led to a correction of the disturbed VLDL metabolism associated with chronic renal failure including a decrease in peripheral lipolysis, a decrease in hepatic TGSR, and an increase in adipose tissue LPL activity. Since cholesterol-rich VLDL are potentially atherogenic, their normalization with insulin treatment in this animal model suggests a viable area of investigation for the prevention of accelerated atherogenesis in chronic renal failure.

References

1   Attman, P.O.; Gustafson, A.: Lipid and carbohydrate metabolism in uremia. Eur. J. clin. Invest. 9: 285–291 (1979).
2   Bagdade, J.D.; Yee, E.; Wilson, D.E.; Shafrir, E.: Hyperlipidemia in renal failure. Studies of plasma lipoproteins, hepatic triglyceride production, and tissue lipoprotein lipase in a chronically uremic rat model. J. Lab. clin. Med. 81: 176–186 (1978).
3   Boudet, J.; Man, N.K.; Pils, P.; Sausse, A.; Funck-Brentano, J.-L.: Experimental chronic renal failure in the rat by electrocoagulation of the renal cortex. Kidney int. 14: 82–86 (1978).
4   Boyd, M.E.; Albright, E.B.; Foster, D.W.; McGarry, J.D.: In vitro reversal of the fasting state of liver metabolism in the rat. Reevaluation of the roles of insulin and glucose. J. clin. Invest. 68: 142–152 (1981).
5   Cattran, D.C.; Fenton, S.S.A.; Wilson, D.R.; Steiner, G.: Defective triglyceride removal in lipemia associated with peritoneal dialysis and hemodialysis. Ann. intern. Med. 85: 29–33 (1976).
6   Cryer, A.: Tissue lipoprotein lipase activity and its action in lipoprotein metabolism. Int. J. Biochem. 13: 525–542 (1981).

7   DeFronzo, R.A.: Pathogenesis of glucose intolerance in uremia. Metabolism 27: 1866–1879 (1978).

8   DeFronzo, R.A.; Alvestrand, A.; Hendler, D.S.: Insulin resistance in uremia. J. clin. Invest. 67: 563–568 (1981).

9   Drueke, T.; Lacour, B.; Roullet, J.-B.; Funck-Brentano, J.-L.: Recent advances in factors that alter lipid metabolism in chronic renal failure. Kidney int. 24: S134–S138 (1983).

10   Duncombe, W.G.: The colorimetric micro-determination of non-esterified fatty acids in plasma. Clinica chim. Acta 9: 122–125 (1964).

11   Goldberg, A.; Sherrard, D.J.; Brunzell, J.D.: Adipose tissue lipoprotein lipase in chronic hemodialysis. Role in plasma triglyceride metabolism. J. clin. Endocr. Metab. 47: 1173–1182 (1978).

12   Havel, R.J.; Eder, H.A.; Bragdon, J.H.: The distribution and chemical composition of ultracentrifugally separated lipoproteins in human serum. J. clin. Invest. 34: 1345–1353 (1955).

13   Heuck, C.C.; Liersch, M.; Ritz, E.; Steigmeier, K.; Wirth, A.; Mehls, O.: Hyperlipoproteinemia in experimental chronic renal insufficiency in the rat. Kidney int. 14: 142–150 (1978).

14   Ibels, L.S.; Reardon, M.F.; Nestel, P.J.: Plasma post-heparin lipolytic activity and triglyceride clearance in uremic and hemodialysis patients and renal allograft recipients. J. Lab. clin. Med. 87: 648–658 (1976).

15   Kauffman, J.M.; Caro, J.F.: Insulin resistance in uremia. Characterization of insulin action, binding and processing in isolated hepatocytes from chronic uremic rats. J. clin. Invest. 71: 698–708 (1983).

16   Li, P.K.; Lee, J.J.; Gillivray, M.H.A.; Schaefer, P.A.; Siegel, J.H.: Direct, fixed-time kinetic assay for β-hydroxybutyrate and aceto-acetate with a centrifugal analyzer or a computer-backed spectrophotometer. Clin. Chem. 12: 1713–1717 (1980).

17   Lowry, O.H.; Rosebrough, N.J.; Farr, A.L.; Randall, R.J.: Protein measurement with the Folin phenol reagent. J. biol. Chem. 193: 265–275 (1951).

18   Mahley, R.W.: Atherogenic hyperlipoproteinemia: the cellular and molecular biology of plasma lipoproteins altered by dietary fat and cholesterol. Med. Clins N. Am. 66: 375–402 (1982).

19   Maloff, B.L.; McCaleb, M.L.; Lockwood, D.H.: Cellular basis of insulin resistance in chronic uremia. Am. J. Physiol. 245: 178–184 (1983).

20   McCaleb, M.L.; Mevorach, R.; Freeman, R.B.; Izzo, M.S.; Lockwood, D.H.: Induction of insulin resistance in normal adipose tissue by uremic human serum. Kidney int. 25: 416–421 (1984).

21   McCormick, K.L.; Widness, J.A.; Susa, J.B.; Schwartz, R.: Effects of chronic hyperinsulinemia on hepatic enzymes involved in lipogenesis and carbohydrate metabolism in the young rat. Biochem. J. 172: 327–331 (1978).

22   Nilsson-Ehle, P.; Schotz, M.C.: A stable radioactive substrate emulsion for assay of lipoprotein lipase. J. Lipid Res. 17: 536–541 (1976).

23   Norbeck, H.E.; Carlson, L.A.: Increased frequency of late pre-β lipoproteins (LPβ) in isolated serum very low density lipoproteins in uremia. Eur. J. clin. Invest. 10: 423–426 (1980).

24   Persson, B.: Lipoprotein lipase activity of human adipose tissue in health and in some diseases with hyperlipidemia as a common feature. Acta med. scand. 193: 457–462 (1973).

25   Rapoport, J.; Aviram, M.; Chaimovitz, C.; Brook, J.G.: Defective high density lipopro-
      tein composition in patients on chronic hemodialysis. A possible mechanism for acceler-
      ated atherosclerosis. New Engl. J. Med. *299:* 1326–1329 (1978).
26   Roullet, J.-B.; Lacour, B.; Yvert, J.-P.; Prat, J.-J.; Drueke, T.: Factors of increase in
      serum triglyceride-rich lipoproteins in uremic rats. Kidney int. *27:* 420–425 (1985).
27   Vydelingum, N.; Drake, R.L.; Etienne, J.; Kissebah, A.H.: Insulin regulation of fat cell
      ribosomes, protein synthesis and lipoprotein lipase. Am. J. Physiol. *245:* E121–E131
      (1983).

Prof. Dr. T. Drueke, Service de Néphrologie, Hôpital Necker, 161, rue de Sèvres,
F-75730 Paris Cedex 15 (France)

Discussion

Quellhorst: Thank you, Dr. Drueke, for your very interesting presentation. Are there any questions or comments?

Massry: These are beautiful data. As I mentioned to you yesterday, I think we can put these data together with the data I presented. By suppressing secondary hyperparathyroidism you are going to overcome the carbohydrate intolerance and probably all these events come together. Indeed, in your precious studies with normal animals in which you produced secondary hyperparathyroidism, you found that you can reproduce these defects. I can tell you that we have almost finished a study in our uremic dogs with and without parathyroid glands and we can show that in uremic dogs with intact parathyroid glands the lipolytic activity is reduced as you have shown. If you do a parathyroidectomy you not only normalise them, but you make them even better than normal. It is possible that, by suppressing parathyroid gland activity, you can achieve the same thing without the need of giving continuous insulin.

Drueke: I was lucky to have found one of the most important underlying causes of decreased lipoprotein lipase (LPL) activity, namely insulin. It is possible that the insufficient insulin secretion, as you suggest, is due to hyperparathyroidism so the real underlying cause might be PTH.

Ardaillou: I would like to ask you two questions and make a brief comment. Firstly, you show that in uremic rats basal plasma insulin is lower than in control rats but usually in uremic patients I thought that plasma insulin levels were greater. Can you comment on this discrepancy between uremic rats and uremic patients? My second question is, would it be possible to study the response of lipoprotein lipase to heparin administration and to see whether or not this response is similar in uremic rats, and in normal rats? Since I am not a specialist I feel it would be of interest.

My comment is that Dr. Etienne in Paris has purified a very good antibody against rat lipoprotein lipase. Perhaps this antibody would be of interest for you to measure more precisely lipoprotein lipase than with the enzyme activity only?

Drueke: Thank you for your questions and your comments. First, the immunoreactive insulin concentration in uremic patients is, in fact, generally increased, not decreased. Generally, one hypothesizes that this increase is not sufficient to overcome the resistance of the uremic organism against insulin. However, several studies show that even in the face of a slightly increased insulin concentration, there are several studies which showed that the insulin secretion rate in uremic patients is decreased, not increased. Thus, there must also be problems with an insufficient renal catabolism of insulin. In uremic rats, generally, lower than normal plasma insulin concentrations are found, except in one or two studies in which the concentrations were high. In our hands, using the particular radioimmunoassay, it is used constantly lower than normal.

As to your second question, the plasma postheparin lipolytic activity which is often evaluated in the human comprises the hepatic and the peripheral triglyceride lipases, i.e. lipoprotein lipases (LPL). However, the peripheral LPL constitutes only a small part, I think, 10–20% of the total triglyceride lipases activity in plasma. If you want to solve the problem of whether LPL or hepatic triglyceride lipase is involved in a particular situation, you have to separate them, after heparin perfusion using different methods, either biochemical or immunological ones, in order to obtain separate evaluations.

Brodde: Could you mimic your insulin effects by oral antidiabetics?

Drueke: We did not try that. It would be an interesting alternative. The relevance of insulin administration to uremic patients may be of small practical value, unless you consider the possibility of administering small amounts of insulin to CAPD patients, who receive large amounts of glucose and in whom accelerated atherosclerosis may be observed in the next 10 or 20 years when these patients can be held long enough on that particular method of treatment. We consider that possibility at the present time.

Quellhorst: Have you any results from the CAPD treatment?

Drueke: No, this is a project. Only very small amounts of insulin should be administered. This should be feasible and I would bet that one might observe an amelioration of plasma LPL just like in the rats and possibly a lesser increase in atherogenic VLDL in circulating blood.

Heidland: I enjoyed your presentation very much. There is one finding that I cannot understand. Insulin is known to work anabolically, but you observed an increase of the BUN level during insulin treatment. We have similar observations in insulin-treated bilateral nephrectomized rats. Surprisingly the BUN levels increased in this group. Now, my question to your experiments concerns the controls. Did you treat non-uremic rats with insulin which might explain the unexpected rise of BUN?

Drueke: No, I have not. In fact, the creatinine levels are the same. So these rats do not have a different renal function as is appreciated by the creatinine value and there must be some other effect which I cannot explain.

Massry: They may eat better.

Drueke: They may eat more and for the same degree of renal insufficiency they could form more urea.

Heidland: The body weight was identical?

Drueke: Yes, the overall body weight was the same, but the fat pad weight was increased in the treated, as compared to the untreated, rats.

Franck: It would be interesting to know in your model what happened to the LDL fraction. You have shown a very low density level only. What happens to the LDL?

Drueke: We have not analyzed the LDL in this study.

Eigler: Do you keep your animals in metabolic cages for 35 days, or how do you manage to get these insulin pumps working for a month or so?

Drueke: Yes, we put them into metabolic cages. It is easier, and the pumps were changed, I think, twice during the whole time. That is, the pump perfusion stability is not long enough to have only one pump intraperitoneally during the whole study.

Subject Index